FATIH AKAY

The Fabric of Universe

Exploring Time and the Wonders of the Cosmos

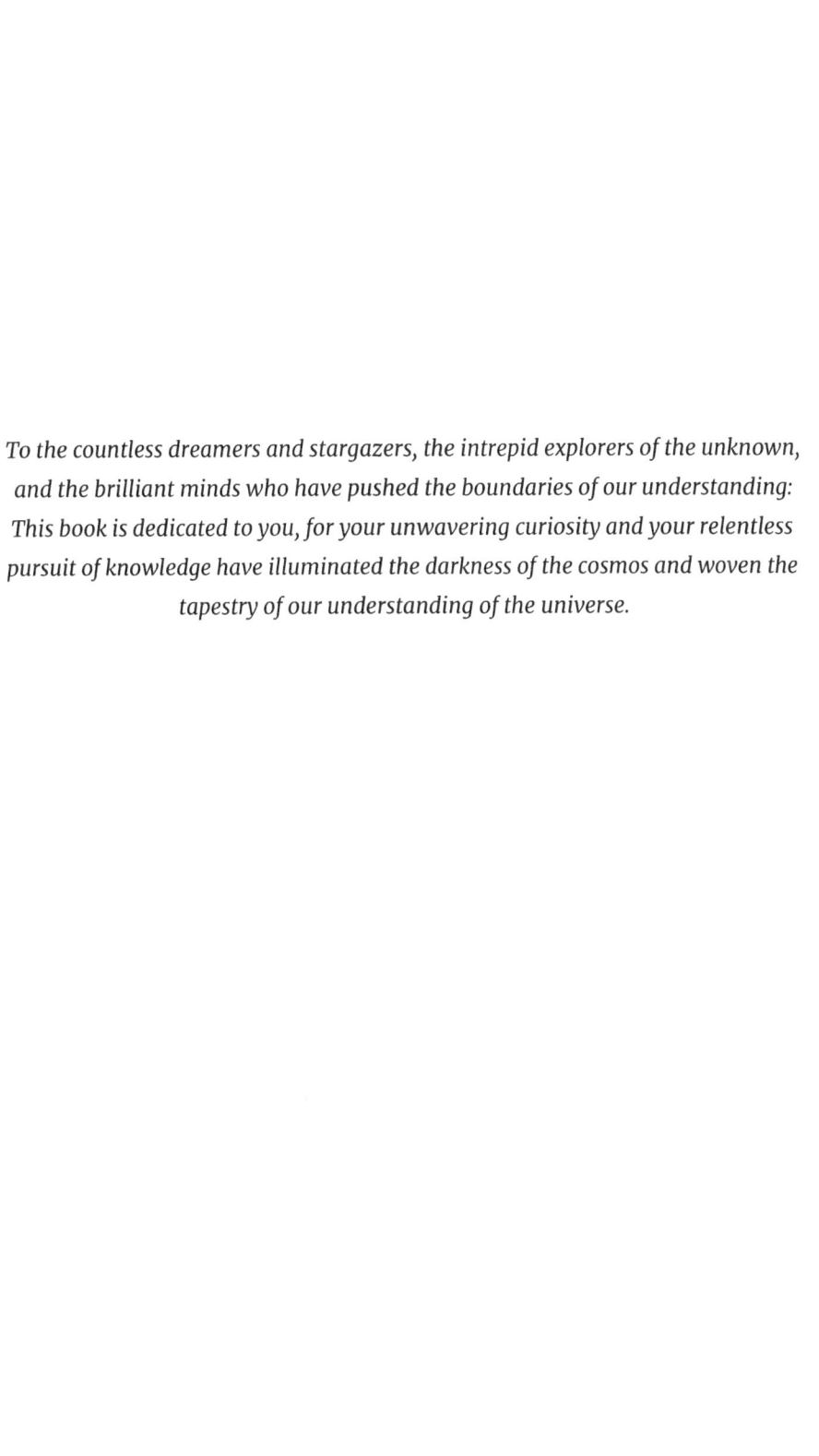

To the countless dreamers and stargazers, the intrepid explorers of the unknown, and the brilliant minds who have pushed the boundaries of our understanding: This book is dedicated to you, for your unwavering curiosity and your relentless pursuit of knowledge have illuminated the darkness of the cosmos and woven the tapestry of our understanding of the universe.

"Look up at the stars and not down at your feet. Try to make sense of what you see, and wonder about what makes the universe exist. Be curious."

-Stephen Hawking

Contents

Foreword

The human story has been, in many ways, a quest to comprehend the vast, seemingly infinite cosmos that surrounds us. As we have gazed up at the stars and pondered our place in the universe, we have been driven by an innate curiosity and a profound desire to unlock the mysteries that lie hidden within the fabric of existence. This book, "The Fabric of Universe: Exploring Time and the Wonders of the Cosmos," is a celebration of that journey, a tribute to the remarkable achievements of the countless individuals who have dedicated their lives to understanding the universe and its many enigmas.

The pages that follow will take you on a voyage through the ages, from the earliest civilizations and their awe-inspired observations of the night sky to the groundbreaking discoveries and theories that have shaped our modern understanding of the cosmos. Along the way, we will encounter the work of visionary thinkers and pioneering scientists whose relentless pursuit of knowledge has forever changed our perception of the universe and our place within it.

As we delve into the intricacies of topics such as the nature of time, the expansion of the universe, and the mysterious phenomena of black holes and wormholes, we will explore not only the astonishing beauty and complexity of the cosmos but also the equally remarkable capacity of the human mind to comprehend it. "The Fabric of Universe" is both a testament to the power of human curiosity and a humble reminder of how much remains to be discovered.

In today's rapidly changing world, it is easy to become preoccupied with the concerns of our daily lives and lose sight of the grand cosmic stage upon which our story unfolds. Yet, as we peer into the depths of the universe and ponder the mysteries of time and space, we are reminded of our shared origins and the interconnected nature of our existence. The study of the cosmos has the power to inspire awe, humility, and wonder, fostering a sense of unity and purpose that transcends the divisions and challenges of our time.

It is my hope that, as you embark on this journey through the pages of "The Fabric of Universe" you will be inspired to look up at the stars and contemplate the wonders of the cosmos, rekindling the spirit of curiosity and exploration that lies at the heart of our human story. May this book serve as a reminder that the mysteries of the universe are as vast and infinite as our capacity to imagine them, and that the journey toward understanding is a journey without end.

Preface

The universe has always been a source of fascination and wonder, inspiring countless generations to explore the mysteries that lie beyond our earthly realm. The more we learn about the cosmos, the more we recognize the intricate tapestry of forces and phenomena that shape the fabric of existence. As a lifelong enthusiast of astronomy and physics, I have been captivated by the incredible story of our evolving understanding of the universe and the enigmatic concept of time that governs it.

"The Fabric of Eternity: Exploring Time and the Wonders of the Cosmos" is the culmination of years of research, contemplation, and passion for the subject. This book aims to offer a comprehensive and accessible exploration of the cosmos and its mysteries, weaving together the threads of history, philosophy, and physics to provide a coherent picture of our place in the universe.

In writing this book, I have endeavored to make the complex ideas and theories that underpin our understanding of the cosmos as clear and accessible as possible, without sacrificing the depth and rigor that the subject demands. The journey begins with a look back at the earliest civilizations and their perceptions of the universe, followed by an examination of the groundbreaking work of scientists and philosophers who have shaped our modern understanding of the cosmos. Along the way, we will delve into the fascinating concepts of the expanding universe, black holes, the nature of time, and the unification of physics, among many others.

I hope that "The Fabric of Universe" will inspire readers to appreciate the awe-inspiring beauty and complexity of the cosmos and to recognize the remarkable achievements of the countless individuals who have dedicated their lives to unraveling its mysteries. As we continue to push the boundaries of our understanding, we are constantly reminded of the vastness of the universe and the infinite potential for discovery that lies within the human spirit.

I would like to express my deepest gratitude to the many mentors, colleagues, friends, and family members who have supported and encouraged me throughout the process of writing this book. Your guidance, insight, and enthusiasm have been invaluable in bringing this project to fruition.

Finally, I would like to thank you, the reader, for embarking on this journey with me. It is my sincere hope that "The Fabric of Universe" will spark your curiosity, ignite your imagination, and awaken a sense of wonder at the marvels of the cosmos. Together, let us continue to explore the mysteries of time and the universe, weaving our own threads into the ever-evolving tapestry of human understanding.

Acknowledgement

Writing "The Fabric of Universe: Exploring Time and the Wonders of the Cosmos" has been an enriching and rewarding experience, made possible by the support, encouragement, and contributions of many individuals. I would like to express my heartfelt gratitude to all those who have played a role in the creation of this book.

First and foremost, I would like to thank my family for their unwavering love, support, and patience throughout this journey. Your belief in me and my work has been a constant source of motivation and inspiration.

To my mentors and professors, who have guided me through my academic journey and ignited my passion for the cosmos, I am deeply grateful. Your wisdom, knowledge, and enthusiasm have shaped my understanding of the universe and fueled my desire to share this knowledge with others.

I would also like to acknowledge my colleagues and peers in the field of astronomy and physics, whose groundbreaking research and discoveries have informed and inspired the content of this book. Your work continues to push the boundaries of our understanding of the universe and its mysteries.

To my friends, who have provided a sounding board for ideas, a shoulder to lean on during challenging moments, and a constant source of encouragement, thank you for your unwavering support and camaraderie.

Finally, I would like to express my gratitude to the readers of "The Fabric of Universe." It is my hope that this book will inspire you to explore the wonders of the cosmos and to appreciate the incredible journey of discovery that has brought us to our current understanding of the universe.

Thank you all for being part of this incredible journey

I

Our Picture of the Universe

1

Introduction to Our Picture of the Universe

Our perception of the universe and our place within it has evolved dramatically over the course of human history. From the earliest days of civilization to the present day, our understanding of the cosmos has been shaped by the development of new technologies, innovative theories, and groundbreaking discoveries. This chapter will serve as an introduction to our picture of the universe, tracing the journey from our earliest observations of the night sky to the complex scientific models of the present day.

The Ancient World's Understanding of Time

The story of humanity's quest to understand the universe begins with the ancient civilizations that populated the Earth thousands of years ago. These early societies observed the motion of celestial bodies like the Sun, Moon, and stars and sought to understand their patterns and significance. In doing so, they laid the foundations for the study of astronomy and our understanding of the cosmos.

For the ancient Egyptians, the universe was a vast, cyclical expanse in which the Earth was a flat surface surrounded by the primeval waters of chaos. Above the Earth, the sky was a great dome, with the stars affixed to its surface. The Sun was a celestial boat that traveled across the sky each day, battling the forces of darkness each night before reemerging in the east at dawn.

The Babylonians, meanwhile, were skilled astronomers who meticulously recorded the positions of celestial bodies, allowing them to predict the movements of the planets, the timing of eclipses, and the changing of the seasons. The Babylonian creation myth, the Enuma Elish, portrayed the cosmos as a battleground between the gods, with the Earth created from the body of the slain goddess Tiamat.

In ancient China, the universe was viewed as a harmonious balance between the opposing forces of yin and yang. The heavens were ruled by the Jade Emperor, while the Earth was supported by a cosmic pillar. Chinese astronomers carefully observed the stars and planets and devised sophisticated systems of astrology to interpret their movements and predict the future.

The Geocentric Model of the Universe

The ancient Greeks made significant contributions to our understanding of the universe, establishing the geocentric model that would dominate Western thought for centuries. The geocentric model posits that the Earth is at the center of the universe, with the celestial bodies, including the Sun, Moon, and stars, revolving around it in perfect circles.

The Greek philosopher Aristotle provided a detailed account of the geocentric model in his work On the Heavens, describing the Earth as a motionless sphere surrounded by a series of concentric celestial spheres, each carrying a planet, the Sun, or the Moon. The outermost sphere, containing the fixed stars, defined the boundary of the universe.

The astronomer and mathematician Claudius Ptolemy further refined the geocentric model in his seminal work, the Almagest. Ptolemy introduced the concepts of epicycles, deferents, and equants to account for the irregular motions of the planets, which could not be adequately explained by simple circular orbits. The Ptolemaic model of the universe was complex but remarkably accurate in predicting the positions of celestial bodies, and it remained the standard view of the cosmos throughout the Middle Ages.

The Copernican Revolution

The geocentric model of the universe began to be challenged in the early 16th century by the work of the Polish astronomer Nicolaus Copernicus. In his book, On the Revolutions of the Celestial Spheres, Copernicus proposed a heliocentric model of the universe, placing the Sun, rather than the Earth, at its center. According to Copernicus, the Earth and other planets revolved around the Sun in circular orbits, with the Moon orbiting the Earth.

Copernicus's revolutionary ideas faced significant opposition from both religious and scientific authorities, who believed that the geocentric model, rooted in tradition and supported by the teachings of Aristotle and Ptolemy, provided a more accurate representation of the universe. However, the seeds of change had been sown, and the heliocentric model would eventually gain traction among astronomers and thinkers in the coming centuries.

Galileo and the Telescope

One of the most influential figures in the shift toward the heliocentric model was the Italian astronomer and physicist Galileo Galilei. In the early 17th century, Galileo made a series of groundbreaking observations using a telescope, a relatively new invention at the time. Through his telescope, Galileo observed mountains and craters on the Moon, sunspots on the Sun, and the phases of Venus, providing further evidence to support the heliocentric model.

Perhaps most significantly, Galileo discovered four large moons orbiting Jupiter, which he named the Galilean moons. This discovery challenged the prevailing view that all celestial bodies must orbit the Earth, demonstrating that other objects in the solar system could have their own satellites. Galileo's observations, published in his book Sidereus Nuncius, brought the heliocentric model to the forefront of scientific discourse and laid the groundwork for future advancements in our understanding of the cosmos.

As the heliocentric model gained acceptance among scientists, the stage was set for a series of discoveries and theories that would revolutionize our picture of the universe. The work of astronomers like Johannes Kepler, who developed the laws of planetary motion, and Sir Isaac Newton, who formulated the laws of motion and universal gravitation, would build upon the foundation laid by Copernicus and Galileo, providing a more comprehensive and accurate understanding of the cosmos.

The Theory of Relativity

In the early 20th century, a new theory emerged that would once again change our perception of the universe: Albert Einstein's theory of relativity. The theory of relativity is composed of two parts: the special theory of relativity, which deals with objects moving at constant speeds (particularly those close to the speed of light), and the general theory of relativity, which deals with gravity and the curvature of spacetime.

Einstein's special theory of relativity introduced the concept that time and space are not absolute but are relative to the observer's frame of reference. This led to the discovery of time dilation and length contraction, phenomena that occur when an object is moving at speeds close to the speed of light.

The general theory of relativity, published by Einstein in 1915, provided a new understanding of gravity as a curvature of spacetime caused by the presence of mass. This theory not only explained the motion of celestial bodies more accurately than Newton's laws but also predicted the existence of black holes and the expansion of the universe.

The development of the theory of relativity marked the beginning of a new era in our understanding of the cosmos, paving the way for the study of topics such as the expanding universe, black holes, and the nature of time itself. As we continue our journey through the history of our picture of the universe, we will explore these fascinating concepts and the scientific advancements that have shaped our current understanding of the cosmos.

From the humble beginnings of ancient civilizations gazing up at the night sky to the complex scientific models of the present day, our picture of the universe has evolved in remarkable ways. As we delve deeper into the mysteries of the cosmos, we continue to push the boundaries of our understanding, unlocking new secrets and inspiring future generations to explore the fabric of eternity.

2

The Ancient World's Understanding of Time

Throughout human history, the concept of time has been of great interest and importance to civilizations across the world. The ancient understanding of time was intimately tied to the natural cycles of the world, such as the changing seasons, the movement of celestial bodies, and the progression of life itself. In this chapter, we will delve into the ancient world's understanding of time, exploring how various civilizations across the globe measured, conceptualized, and utilized time in their daily lives and beliefs.

· The Sumerians and the Invention of the Sexagesimal System

The Sumerians, one of the world's oldest known civilizations, made significant advancements in the understanding and measurement of time. They developed the sexagesimal system, a numerical system based on the number 60, which is still in use today for measuring time, angles, and geographic coordinates. This system allowed the Sumerians to divide the day into 24 hours, each hour into 60 minutes, and each minute into 60 seconds.

The Sumerians also created one of the earliest known lunar calendars, using the cycles of the moon to measure the passage of time. Their calendar consisted of 12 lunar months, each approximately 29.5 days long. To reconcile the difference between the lunar year and the solar year, the Sumerians added an extra month every few years, a practice known as intercalation.

· The Egyptians and the Solar Calendar

The ancient Egyptians were among the first to develop a solar calendar, based on the apparent movement of the sun across the sky. They divided the year into 365 days, with each day beginning at sunrise. The Egyptian calendar consisted of 12 months, each with 30 days, and an additional five days at the end of the year, known as the epagomenal days. These additional days were dedicated to celebrating the births of five gods and goddesses and were considered outside the normal calendar year.

The Egyptians also developed sophisticated sundials and water clocks to measure time throughout the day. Sundials were used to measure the movement of the sun, while water clocks relied on the steady flow of water to mark the passage of time. These devices allowed the Egyptians to measure time more accurately than ever before, enabling them to better plan their agricultural activities and religious ceremonies.

· The Mesoamerican Calendars: The Maya and the Aztecs

In Mesoamerica, both the Maya and Aztec civilizations developed complex and highly accurate calendar systems. The Maya calendar was composed of two interlocking cycles: the Tzolk'in, a 260-day sacred calendar, and the Haab', a 365-day solar calendar. The Tzolk'in was made up of 20 day-names, each associated with a specific god or goddess, and 13 numbers, which created a unique combination for each day of the cycle. The Haab' consisted of 18 months, each with 20 days, and a five-day period called Wayeb' at the end of the year, which was considered a time of uncertainty and potential danger.

The Aztecs similarly used two interlocking calendar systems: the Xiuhpohualli, a 365-day solar calendar, and the Tonalpohualli, a 260-day sacred calendar. The Xiuhpohualli was divided into 18 months, each with 20 days, and a five-day period called Nemontemi, which was also considered a dangerous and inauspicious time. The Tonalpohualli, like the Maya Tzolk'in, was made up of 20 day-names and 13 numbers, creating a unique combination for each day of the cycle.

These Mesoamerican calendars were central to the religious and social lives of the Maya and Aztec people, with specific days considered auspicious or inauspicious for various activities, such as farming, building, and conducting ceremonies. Both civilizations also developed sophisticated astronomical knowledge and made accurate predictions of celestial events, such as eclipses, using their calendars.

1. The Ancient Chinese Calendar and Timekeeping

The ancient Chinese developed a complex lunar-solar calendar system, which combined the lunar cycles with the solar year. This system, known as the lunisolar calendar, was used to track time, agricultural activities, and religious observances. The Chinese calendar consisted of 12 or 13 lunar months, each beginning with the sighting of a new moon, with intercalation used to keep the calendar in alignment with the solar year.

In addition to the lunisolar calendar, the ancient Chinese developed sophisticated timekeeping devices, such as water clocks and incense clocks. Water clocks used the steady flow of water to measure the passage of time, while incense clocks relied on the consistent burning rate of incense sticks. These devices enabled the Chinese to measure time with great precision, allowing them to plan their daily activities and important events accordingly.

· **The Greek Concept of Time: Chronos and Kairos**

In ancient Greece, the concept of time was embodied by two distinct deities: Chronos and Kairos. Chronos represented linear, sequential time, which could be measured and quantified. The Greeks developed various timekeeping devices, such as sundials and water clocks, to measure Chronos, and they used a lunar calendar, similar to the one used by the Sumerians, to track the passage of time.

Kairos, on the other hand, represented the opportune moment, the qualitative aspect of time that could not be measured. Kairos was associated with critical junctures, when the right action or decision needed to be made. This concept of time was central to Greek philosophy, particularly in the works of Plato and Aristotle, who saw time as a fundamental aspect of human existence, intimately connected to the concepts of change, motion, and eternity.

· The Roman Calendar and the Julian Reform

The ancient Romans adopted and adapted the calendar systems of the civilizations they encountered, particularly those of the Greeks and the Egyptians. The Roman calendar was originally a lunar calendar, consisting of 10 months and 304 days, with an additional 61 days in the winter period. However, this calendar quickly fell out of alignment with the solar year, leading to the introduction of the Julian calendar by Julius Caesar in 45 BCE.

The Julian calendar was a solar calendar, based on the Egyptian model, with 12 months and 365 days, and a leap year every four years. This calendar was more accurate than its predecessors and became the dominant calendar system in the Western world for over 1,600 years, until it was replaced by the Gregorian calendar in 1582.

The ancient world's understanding of time was shaped by the natural cycles of the world and the movement of celestial bodies, with each civilization developing unique calendar systems and timekeeping devices to measure and conceptualize time. These ancient concepts of time have had a profound and lasting impact on our modern understanding of time, influencing our calendar systems, timekeeping devices, and the very way we perceive the passage of time.

3

The Geocentric Model of the Universe

The geocentric model of the universe, also known as the Ptolemaic system, was the dominant astronomical paradigm for over a thousand years. This model placed Earth at the center of the universe, with all celestial bodies revolving around it in perfect circles. In this chapter, we will delve into the origins of the geocentric model, its key proponents and critics, and its eventual decline in the face of the heliocentric model proposed by Copernicus.

· **The Origins of the Geocentric Model**

The idea of a geocentric universe can be traced back to ancient Mesopotamia, where the Babylonians observed the motion of celestial bodies and concluded that the Earth must be at the center of the universe. This idea was further developed by ancient Greek philosophers, such as Parmenides and Plato, who believed that the heavens were perfect and unchanging, and thus could only be explained by a model in which celestial bodies moved in perfect circles around the Earth.

· **The Eudoxan Model**

Eudoxus of Cnidus, a Greek mathematician and astronomer, was one of the first to develop a detailed geocentric model of the universe. His model featured concentric spheres, with the Earth at the center, and the planets, the sun, and the moon all embedded in their respective spheres. Each sphere would rotate on its axis, creating the appearance of the celestial bodies moving across the sky. Eudoxus' model was an important precursor to the Ptolemaic system, and it was widely influential in its time.

· The Aristotelian Model

Aristotle, one of the most famous philosophers of ancient Greece, further refined the geocentric model. He argued that the universe was divided into two realms: the sublunary realm, which consisted of the Earth and the surrounding air and was subject to change and decay, and the celestial realm, which was perfect, unchanging, and governed by different laws. In his model, the celestial bodies were embedded in crystalline spheres that rotated around the Earth. Aristotle's model was highly influential and shaped the way people thought about the universe for centuries.

· Claudius Ptolemy and the Ptolemaic System

Claudius Ptolemy, a Greek-Roman astronomer who lived in the 2nd century CE, was responsible for the most comprehensive and sophisticated geocentric model of the universe. His model, known as the Ptolemaic system, incorporated the idea of epicycles, smaller circles along which the celestial bodies moved as they orbited the Earth. The addition of epicycles allowed Ptolemy's model to better account for the irregular motion of planets, particularly their retrograde motion, when they appear to move backward in the sky.

Ptolemy's model was laid out in his work, the Almagest, which became the authoritative text on astronomy for over a thousand years. The Ptolemaic system was widely accepted by scholars, theologians, and the general public, as it aligned well with the religious teachings of the time, which placed Earth and humanity at the center of God's creation.

· The Islamic Golden Age and the Preservation of the Geocentric Model

During the Islamic Golden Age, between the 8th and 13th centuries, Islamic scholars translated and preserved the works of Greek and Roman astronomers, including Ptolemy's Almagest. They also made significant contributions to the geocentric model, refining its mathematics and developing improved astronomical instruments, such as the astrolabe, for making precise observations of celestial bodies.

Notable Islamic astronomers, such as Al-Battani, Al-Farghani, and Al-Sufi, made significant contributions to the understanding of celestial motions and refined the Ptolemaic system. Although some Islamic astronomers, such as Alhazen, began to question the geocentric model, it remained the dominant paradigm in the Islamic world throughout the Middle Ages.

· The European Renaissance and the Revival of the Geocentric Model

During the European Renaissance, the rediscovery of ancient Greek and Roman texts, including Ptolemy's Almagest, sparked a renewed interest in the geocentric model. Astronomers like Regiomontanus and Johannes Müller (also known as Regiomontanus) worked to further refine the Ptolemaic system, developing new mathematical techniques and observational instruments.

At the same time, however, the geocentric model began to face increasing criticism from some astronomers, who pointed out its growing inconsistencies with observed celestial motions. One of the most significant challenges to the geocentric model came from Nicolaus Copernicus, a Polish astronomer who proposed a heliocentric model of the universe.

· Nicolaus Copernicus and the Heliocentric Model

Nicolaus Copernicus, in his seminal work, De revolutionibus orbium coelestium (On the Revolutions of the Heavenly Spheres), proposed a heliocentric model of the universe, in which the Sun was at the center, and the Earth and other planets revolved around it in circular orbits. Copernicus' model offered a simpler explanation for the observed motions of celestial bodies, including the retrograde motion of planets, and eliminated the need for epicycles.

Although Copernicus' heliocentric model faced strong opposition from religious authorities and some astronomers, it gradually gained acceptance among the scientific community. Astronomers such as Tycho Brahe, Johannes Kepler, and Galileo Galilei made observations and developed mathematical models that further supported the heliocentric model and challenged the geocentric paradigm.

· Tycho Brahe and the Tychonic System

Tycho Brahe, a Danish astronomer, made some of the most accurate and comprehensive astronomical observations of his time, before the invention of the telescope. Although Brahe did not fully embrace the heliocentric model, he proposed a hybrid system, known as the Tychonic system, in which the Earth remained stationary at the center of the universe, with the Sun and Moon orbiting around it, and the other planets orbiting around the Sun. The Tychonic system represented a significant step away from the geocentric model and paved the way for the eventual acceptance of the heliocentric model.

· Johannes Kepler and the Laws of Planetary Motion

Johannes Kepler, a German astronomer and mathematician, was a key figure in the transition from the geocentric to the heliocentric model. Using Tycho Brahe's precise observational data, Kepler developed his three laws of planetary motion, which described the motion of planets in elliptical orbits around the Sun. Kepler's laws provided strong mathematical support for the heliocentric model and further undermined the geocentric model, which could not account for the observed elliptical orbits without introducing additional epicycles and complexities.

· Galileo Galilei and the Telescope

Galileo Galilei, an Italian astronomer, mathematician, and philosopher, played a pivotal role in the decline of the geocentric model. Using the newly invented telescope, Galileo made groundbreaking observations of celestial bodies, such as the moons of Jupiter, the phases of Venus, and the mountains and craters on the Moon. These observations provided strong evidence for the heliocentric model and directly contradicted the geocentric model's predictions.

Despite facing opposition from religious authorities and being placed under house arrest, Galileo's discoveries and writings had a profound impact on the scientific community. His observations and arguments in favor of the heliocentric model helped to shift the dominant paradigm away from the geocentric model and set the stage for the modern understanding of the universe.

1. Isaac Newton and the Laws of Motion and Universal Gravitation

Isaac Newton, an English mathematician, physicist, and astronomer, made groundbreaking contributions to the understanding of celestial motion and the laws that govern the universe. His laws of motion and universal gravitation provided a comprehensive framework for understanding the motion of celestial bodies, including the planets, moons, and comets.

Newton's laws not only supported the heliocentric model but also explained the elliptical orbits observed by Kepler, as well as other phenomena, such as the tides on Earth. His work, particularly the Principia Mathematica, marked the end of the geocentric model's dominance and established the heliocentric model as the foundation of modern astronomy.

1. The Legacy of the Geocentric Model

Although the geocentric model has long been superseded by the heliocentric model, it remains an important historical milestone in the development of our understanding of the universe. For over a thousand years, the geocentric model shaped the way people perceived the cosmos and their place within it. Its eventual decline and replacement by the heliocentric model marked a significant shift in human thought, as it challenged the long-held belief that Earth and humanity were at the center of the universe.

The geocentric model's decline also serves as a powerful reminder of the importance of questioning established paradigms and being open to new ideas in the pursuit of knowledge. The transition from the geocentric to the heliocentric model represents one of the most profound shifts in scientific thought and laid the groundwork for the modern scientific method, which values empirical evidence and mathematical rigor over dogma and tradition.

The geocentric model of the universe was a product of its time, reflecting the best available knowledge and observations of the ancient and medieval worlds. Although it was ultimately replaced by the more accurate heliocentric model, the geocentric model remains an important chapter in the history of astronomy and the development of our understanding of the cosmos. The lessons learned from the geocentric model's rise and fall continue to inform our approach to scientific inquiry and our ongoing quest to unravel the mysteries of the universe.

4

The Copernican Revolution

The Copernican Revolution refers to the paradigm shift in astronomy that occurred during the 16th and 17th centuries when the heliocentric model proposed by Nicolaus Copernicus replaced the long-standing geocentric model. This revolution had a profound impact on not only the field of astronomy but also on human thought, the scientific method, and our understanding of the universe. In this chapter, we will explore the key events, discoveries, and figures that contributed to the Copernican Revolution.

· Nicolaus Copernicus and the Heliocentric Model

Nicolaus Copernicus, a Polish astronomer, was the first to propose a comprehensive heliocentric model of the universe in his seminal work, De revolutionibus orbium coelestium (On the Revolutions of the Heavenly Spheres). Copernicus' model placed the Sun at the center of the universe, with the Earth and other planets orbiting around it in circular paths. This revolutionary idea challenged the long-held belief in the geocentric model and paved the way for the scientific discoveries that would follow.

· The Heliocentric Model's Reception

Initially, the reception of Copernicus' heliocentric model was mixed. While some astronomers and mathematicians recognized the elegance and simplicity of the model, others, particularly religious authorities, viewed it as a threat to the traditional understanding of the cosmos and the position of Earth and humanity within it. Nevertheless, Copernicus' work laid the foundation for the discoveries and innovations that would eventually lead to the widespread acceptance of the heliocentric model.

· Tycho Brahe's Observations

Tycho Brahe, a Danish nobleman and astronomer, made some of the most accurate and comprehensive astronomical observations of his time, before the invention of the telescope. Although Brahe did not fully embrace the heliocentric model, his observations of celestial bodies, particularly the positions and motions of the planets, provided essential data for later astronomers to refine and test the heliocentric model.

· Johannes Kepler and the Laws of Planetary Motion

Johannes Kepler, a German astronomer and mathematician, used Tycho Brahe's precise observational data to develop his three laws of planetary motion, which described the motion of planets in elliptical orbits around the Sun. Kepler's laws provided strong mathematical support for the heliocentric model and further undermined the geocentric model, which could not account for the observed elliptical orbits without introducing additional epicycles and complexities.

· Galileo Galilei and the Telescope

Galileo Galilei, an Italian astronomer, mathematician, and philosopher, played a pivotal role in the Copernican Revolution. Using the newly invented telescope, Galileo made groundbreaking observations of celestial bodies, such as the moons of Jupiter, the phases of Venus, and the mountains and craters on the Moon. These observations provided strong evidence for the heliocentric model and directly contradicted the geocentric model's predictions.

Galileo's discoveries and his advocacy for the heliocentric model brought him into conflict with the Catholic Church, which ultimately led to his trial and house arrest. Despite this opposition, Galileo's work had a profound impact on the scientific community and helped to shift the dominant paradigm away from the geocentric model.

· The Trial of Galileo and the Church's Opposition

The Catholic Church, which held significant power and influence in the 17th century, viewed the heliocentric model as a challenge to its religious teachings, which placed Earth at the center of the universe. Galileo's trial and subsequent house arrest served as a cautionary tale for other astronomers and scientists, who were forced to tread carefully when discussing the heliocentric model and its implications. However, despite the Church's opposition, the evidence in favor of the heliocentric model continued to accumulate, and the scientific community gradually moved toward its acceptance.

· Isaac Newton and the Laws of Motion and Universal Gravitation

Isaac Newton, an English mathematician, physicist, and astronomer, made groundbreaking contributions to the understanding of celestial motion and the laws that govern the universe. His laws of motion and universal gravitation provided a comprehensive framework for understanding the motion of celestial bodies, including the planets, moons, and comets.

Newton's laws not only supported the heliocentric model but also explained the elliptical orbits observed by Kepler, as well as other phenomena, such as the tides on Earth. His work, particularly the Principia Mathematica, marked the end of the geocentric model's dominance and established the heliocentric model as the foundation of modern astronomy.

· The Impact of the Copernican Revolution on the Scientific Method

The Copernican Revolution had a profound impact on the development of the scientific method. The transition from the geocentric to the heliocentric model highlighted the importance of empirical evidence, mathematical rigor, and skepticism in the pursuit of scientific knowledge. The Copernican Revolution also demonstrated the value of questioning long-held beliefs and dogmas and being open to new ideas and discoveries.

· **The Cultural and Philosophical Implications of the Copernican Revolution**

The Copernican Revolution had far-reaching cultural and philosophical implications. By displacing Earth from the center of the universe, the heliocentric model challenged the long-held belief in the special status of Earth and humanity within the cosmos. This shift in perspective contributed to the development of a more secular worldview and paved the way for the Enlightenment, a period of intellectual and cultural change that emphasized reason, skepticism, and the power of human understanding.

· **The Legacy of the Copernican Revolution**

The Copernican Revolution marks a turning point in human history, as it ushered in a new era of scientific discovery and fundamentally changed our understanding of the universe. The heliocentric model and the discoveries that followed have shaped modern astronomy and continue to inform our ongoing quest to unravel the mysteries of the cosmos.

Moreover, the Copernican Revolution serves as a powerful reminder of the importance of challenging established paradigms and being open to new ideas and discoveries. As we continue to explore the universe and push the boundaries of our knowledge, the lessons learned from the Copernican Revolution continue to inspire and guide us in our pursuit of a deeper understanding of the cosmos.

The Copernican Revolution represents a transformative moment in the history of astronomy and human thought. The shift from the geocentric to the heliocentric model not only revolutionized our understanding of the universe but also laid the groundwork for the development of the modern scientific method and the broader cultural and philosophical changes that would follow. The Copernican Revolution serves as a testament to the power of human curiosity, ingenuity, and our unyielding drive to uncover the mysteries of the universe.

5

Galileo and the Telescope

Galileo Galilei, an Italian astronomer, mathematician, and philosopher, played a pivotal role in the history of astronomy and the Copernican Revolution. His use of the telescope to observe celestial bodies and his staunch defense of the heliocentric model transformed our understanding of the universe and laid the groundwork for the modern era of astronomy. In this chapter, we will explore the life of Galileo, his groundbreaking discoveries, and the impact of the telescope on the field of astronomy.

- **Galileo's Early Life and Education**

Galileo Galilei was born in Pisa, Italy, on February 15, 1564. He was the first of six children born to Vincenzo Galilei, a musician and composer, and Giulia Ammannati, a member of an influential Florentine family. Galileo initially studied medicine at the University of Pisa, but his interests quickly turned to mathematics and natural philosophy. He left the university without obtaining a degree but continued to study mathematics and physics independently.

- **Galileo's Academic Career and Early Achievements**

In 1589, Galileo was appointed as a lecturer in mathematics at the University of Pisa, where he began his work on the physics of motion. He conducted experiments on the motion of falling bodies, which contradicted the widely accepted Aristotelian notion that heavier objects fall faster than lighter ones. Galileo's work in this area laid the foundation for the development of classical mechanics and the laws of motion that would later be formulated by Isaac Newton.

In 1592, Galileo moved to the University of Padua, where he continued to teach mathematics and further developed his ideas on motion and mechanics. During this period, Galileo also began to study astronomy and made several important discoveries, such as the isochronism of pendulums and the principle of inertia.

· The Invention of the Telescope

Although Galileo did not invent the telescope, he was the first to apply it to astronomical observations. The telescope was originally invented in the Netherlands around 1608, and Galileo learned of its existence in 1609. He quickly built his own version of the telescope, improving upon the original design, and began to make observations of the night sky.

· Galileo's Observations of the Moon

One of Galileo's first significant astronomical observations using the tele-scope was the discovery of mountains and craters on the Moon. Until this time, the Moon was thought to be a perfectly smooth, unblemished sphere, in accordance with the Aristotelian view of the heavens. Galileo's observations revealed that the Moon's surface was irregular and uneven, similar to Earth's terrain, challenging the prevailing notion of the Moon's perfection.

· The Discovery of the Moons of Jupiter

In 1610, Galileo discovered four moons orbiting Jupiter, which are now known as the Galilean moons: Io, Europa, Ganymede, and Callisto. This discovery provided strong evidence for the Copernican heliocentric model, as it demonstrated that not all celestial bodies orbited Earth, contrary to the geocentric view. The discovery of Jupiter's moons also supported the idea that the universe was much larger and more complex than previously thought.

· The Observation of the Phases of Venus

Another crucial observation made by Galileo was the observation of the phases of Venus. Through his telescope, he observed that Venus exhibited phases similar to the Moon, which could only be explained if Venus orbited the Sun rather than Earth. This observation further supported the heliocentric model and contradicted the geocentric model, which could not account for the observed phases of Venus without introducing additional complexities.

· Galileo's Advocacy for the Heliocentric Model

Armed with these groundbreaking discoveries, Galileo became a staunch advocate for the Copernican heliocentric model. He wrote and published several works in support of the heliocentric model, including The Starry Messenger (Sidereus Nuncius) in 1610 and Letters on Sunspots in 1613. These publications not only detailed his observations but also provided strong arguments in favor of the heliocentric model, significantly contributing to the eventual acceptance of the heliocentric view.

· Conflict with the Catholic Church

Galileo's advocacy for the heliocentric model brought him into conflict with the Catholic Church, which held significant power and influence during his time. The Church supported the geocentric model, as it was consistent with biblical passages that seemed to place Earth at the center of the universe. Galileo's views were seen as heretical, and he was warned by the Church to stop promoting the heliocentric model.

In 1632, Galileo published his most famous work, Dialogue Concerning the Two Chief World Systems, in which he compared the geocentric and heliocentric models. The book was seen as a direct challenge to the Church's authority, and Galileo was summoned to Rome to stand trial before the Inquisition in 1633.

· The Trial of Galileo and House Arrest

During his trial, Galileo was found guilty of heresy for supporting the heliocentric model and was forced to recant his views publicly. He was sentenced to house arrest for the remainder of his life and was forbidden from publishing any further works. Despite his confinement, Galileo continued to conduct research and make important contributions to science, including the study of motion and the development of an early version of the pendulum clock.

· The Legacy of Galileo and the Telescope

Galileo's use of the telescope to make groundbreaking astronomical observations and his unwavering support for the heliocentric model transformed our understanding of the universe and laid the groundwork for the modern era of astronomy. His discoveries and innovations inspired future generations of astronomers and scientists and played a crucial role in the eventual acceptance of the heliocentric model.

The telescope itself revolutionized the field of astronomy, allowing for the observation of celestial bodies and phenomena that were previously invisible to the naked eye. The telescope has continued to evolve over the centuries, leading to increasingly powerful instruments, such as the Hubble Space Telescope and the upcoming James Webb Space Telescope. These instruments have allowed us to explore the universe in unprecedented detail, vastly expanding our knowledge of the cosmos and our place within it.

Galileo Galilei and his use of the telescope had a profound impact on the history of astronomy and our understanding of the universe. His groundbreaking discoveries, unyielding support for the heliocentric model, and the development of the telescope as a tool for astronomical observation have left an enduring legacy that continues to shape the field of astronomy and our ongoing quest to unravel the mysteries of the cosmos.

II

Space and Time

6

Introduction to Space and Time

The concepts of space and time are fundamental to our understanding of the universe and our place within it. From the earliest human civilizations to the cutting-edge theories of modern physics, the study of space and time has shaped our understanding of reality and led to countless scientific discoveries and technological advancements. In this chapter, we will delve into the fascinating world of space and time, exploring their nature, history, and significance in the context of the universe.

· **Early Human Conceptions of Space and Time**

The earliest human civilizations developed their own unique understandings of space and time based on their observations of the natural world and the heavens. For instance, ancient cultures such as the Egyptians, Babylonians, and Mayans developed sophisticated calendars and astronomical systems that allowed them to track the passage of time and predict celestial events such as eclipses and the movements of the planets.

· **The Geocentric Model and the Influence of Aristotle**

In ancient Greece, the philosopher Aristotle played a significant role in shaping the understanding of space and time. He proposed a geocentric model of the universe, with Earth at its center and the celestial bodies orbiting around it in perfect spheres. This model persisted for nearly two millennia and influenced the development of both astronomy and physics.

Aristotle also believed that time was a measure of change and that it was inherently tied to motion. He argued that time could not exist independently of change, and that without change, there would be no time. This view of time as a measure of change became a cornerstone of Western philosophy and influenced later thinkers such as St. Augustine, who saw time as a divine creation that marked the unfolding of God's plan.

· The Copernican Revolution and the Heliocentric Model

The Copernican Revolution, which occurred during the 16th and 17th centuries, marked a significant shift in our understanding of space and time. As discussed in earlier chapters, Nicolaus Copernicus proposed a heliocentric model of the universe, with the Sun at its center and the Earth and other planets orbiting around it. This model challenged the long-standing geocentric model and sparked a revolution in astronomy that fundamentally changed our view of the universe.

The heliocentric model had far-reaching implications for the study of space and time. For instance, it suggested that the universe was much larger than previously believed, and that Earth was not the center of the cosmos. This new perspective led to a deeper appreciation of the vastness of space and the insignificance of our place within it.

· Isaac Newton and the Laws of Motion and Universal Gravitation

Isaac Newton's groundbreaking work in the late 17th century laid the foundation for classical mechanics and our modern understanding of space and time. His laws of motion and universal gravitation provided a comprehensive framework for understanding the motion of celestial bodies, as well as objects on Earth. Newton's work also introduced the concept of absolute space and time, which he believed were independent entities that existed independently of any observer or physical object.

· The Theory of Relativity and the Revolution in Space and Time

In the early 20th century, Albert Einstein revolutionized our understanding of space and time with his theory of relativity. This theory, which comprises both the special and general theories of relativity, demonstrated that space and time were not independent entities but rather interconnected aspects of a four-dimensional spacetime.

Special relativity, which deals with objects moving at constant speeds relative to each other, introduced the concept of time dilation, whereby the passage of time is perceived differently by observers moving at different velocities. This theory also introduced the idea of length contraction, whereby objects appear to be shorter when viewed by an observer in motion relative to the object.

General relativity, which deals with gravity and the curvature of spacetime, showed that gravity is not a force that acts across space, but rather a result of the curvature of spacetime caused by the presence of massive objects. This theory also predicted the existence of black holes and the phenomenon of gravitational time dilation.

Einstein's theory of relativity fundamentally changed our understanding of space and time, and its predictions have been verified by numerous experiments and observations. It has also led to numerous technological advancements, such as GPS, which relies on the precise measurements of time and the effects of relativity on the behavior of signals in space.

· Quantum Mechanics and the Uncertainty Principle

In the early 20th century, the development of quantum mechanics revolutionized our understanding of space and time on the microscopic scale. This theory, which deals with the behavior of particles at the atomic and subatomic level, introduced the concept of wave-particle duality, whereby particles can exhibit both wave-like and particle-like behavior.

Quantum mechanics also introduced the uncertainty principle, which states that the position and momentum of a particle cannot be measured with absolute precision at the same time. This principle has profound implications for our understanding of space and time, as it suggests that the very act of observing particles affects their behavior and changes the nature of reality.

· The Search for a Unified Theory

Despite the enormous progress made in the study of space and time, there is still much that we do not understand. One of the biggest challenges facing modern physics is the development of a unified theory that can reconcile the principles of quantum mechanics and relativity and provide a comprehensive understanding of the nature of the universe.

Numerous theories have been proposed, such as string theory and loop quantum gravity, but none have yet been conclusively proven or accepted by the scientific community. The search for a unified theory remains one of the most exciting and challenging frontiers in the study of space and time.

The concepts of space and time have played a central role in the development of human understanding and scientific progress. From the earliest human civilizations to the cutting-edge theories of modern physics, our study of space and time has transformed our understanding of the universe and led to countless scientific discoveries and technological advancements. As we continue to explore these concepts, we will undoubtedly uncover new mysteries and challenge our understanding of reality.

7

Isaac Newton and the Principia Mathematica

Sir Isaac Newton (1642-1727) was an English physicist, mathematician, astronomer, and author who is widely considered one of the most influential scientists in history. His groundbreaking work, the Philosophiæ Naturalis Principia Mathematica (Mathematical Principles of Natural Philosophy), published in 1687, laid the foundation for modern physics and established the field of classical mechanics. In this paper, we will explore the life of Isaac Newton, the historical context of the Principia Mathematica, and its lasting impact on the scientific world.

Early Life and Education of Isaac Newton

Isaac Newton was born on January 4, 1643, in Woolsthorpe, Lincolnshire, England. His father, a farmer, died three months before Newton's birth. His mother remarried, and Newton was raised by his grandparents until the age of 12 when his mother returned, following the death of her second husband. Despite these difficult circumstances, Newton displayed exceptional intelligence from an early age.

Newton attended the King's School in Grantham, where he excelled in both academics and craftsmanship. In 1661, he entered Trinity College, Cambridge, to study law. However, his interests quickly shifted towards mathematics and natural philosophy (the precursor to modern science). At Cambridge,

Newton was exposed to the works of Aristotle, Descartes, Galileo, and Kepler, which greatly influenced his intellectual development.

The Emergence of the Principia Mathematica

Newton's most significant contributions to science began to take shape in the mid-1660s. While at Cambridge, he developed a series of mathematical methods that would eventually become calculus (simultaneously and independently developed by German mathematician Gottfried Wilhelm Leibniz). Additionally, he made significant discoveries in optics, including the realization that white light is a combination of all colors in the spectrum.

In 1665, the Great Plague forced Cambridge to close, and Newton returned to his family's estate in Woolsthorpe. During this period, he continued his studies and conducted experiments in isolation. This time, known as his annus mirabilis or "year of wonders," proved to be incredibly productive. It was during this time that Newton developed his theory of gravitation, inspired by the famous incident of an apple falling from a tree.

Newton's work on gravitation remained largely unknown until 1679 when he began corresponding with the British astronomer and mathematician Edmond Halley. Halley, impressed by Newton's work, encouraged him to publish his findings. In 1684, Newton began working on the Principia Mathematica, which he completed in just 18 months. Financially supported by Halley, the Principia Mathematica was published in 1687.

The Principia Mathematica and the Three Laws of Motion

The Principia Mathematica is composed of three books. The first book, "De motu corporum," deals with the laws of motion and the mathematical principles governing the movement of bodies. It is in this book that Newton introduces his three famous laws of motion:

1. An object at rest will stay at rest, and an object in motion will stay in motion with a constant velocity, unless acted upon by an external force. This is known as the law of inertia.
2. The force acting on an object is equal to the mass of the object multiplied by its acceleration ($F=ma$). This is the law of force and acceleration.
3. For every action, there is an equal and opposite reaction. This is the law

of action and reaction.

These laws formed the basis of classical mechanics and have been fundamental in the development of physics ever since.

The second book, "De motu corporum libri secundus," focuses on the motion of fluids and the forces acting on them. This section of the Principia Mathematica further established Newton's expertise in the field of fluid dynamics.

The third book, "De mundi systemate," delves into the application of the principles laid out in the first two books to the motion of celestial bodies. Here, Newton presents his law of universal gravitation, which states that every particle in the universe attracts every other particle with a force that is proportional to the product of their masses and inversely proportional to the square of the distance between them ($F = G * (m_1 * m_2) / r^2$). This law explained the observed motions of the planets, moons, and comets, unifying the celestial and terrestrial realms under one set of physical laws.

The Impact of the Principia Mathematica

The publication of the Principia Mathematica had a profound impact on the scientific community. It challenged the Aristotelian worldview, which had dominated Western thought for nearly two millennia, and replaced it with a new framework that united the terrestrial and celestial realms under a single set of mathematical principles. This paradigm shift marked the beginning of the Scientific Revolution and the Age of Enlightenment, as thinkers across Europe began to adopt a more rigorous, empirical approach to understanding the natural world.

The Principia Mathematica also had a lasting impact on the field of mathematics. Newton's development of calculus allowed scientists and mathematicians to model complex phenomena more accurately and to solve previously intractable problems. The methods and concepts introduced by Newton continue to be used in various fields of science and engineering today, including physics, astronomy, and engineering.

Newton's ideas and discoveries also influenced many other notable scientists, including Albert Einstein. The Principia Mathematica's principles played a crucial role in shaping Einstein's theories of special and general relativity, which further revolutionized our understanding of space, time, and gravity.

Legacy and Conclusion

Isaac Newton's life and work represent a critical turning point in the history of science. His discoveries in mathematics, physics, and astronomy laid the foundation for the modern scientific method and established the field of classical mechanics, which would dominate scientific thought for centuries.

The Principia Mathematica, in particular, stands as a testament to Newton's genius and the transformative power of his ideas. It marked the beginning of a new era in scientific thought, one that would eventually lead to a deeper understanding of the universe and our place within it. Today, the Principia Mathematica is considered one of the most important works in the history of science, and its influence continues to be felt in nearly every branch of scientific inquiry.

Sir Isaac Newton's life and his seminal work, the Philosophiæ Naturalis Principia Mathematica, have had an immeasurable impact on the scientific world. His groundbreaking contributions to mathematics, physics, and astronomy continue to shape our understanding of the universe and inspire generations of scientists and thinkers. The legacy of Isaac Newton and the Principia Mathematica will forever be remembered as a pivotal moment in human history when the mysteries of the cosmos began to be unlocked through the power of mathematics and the scientific method.

8

The Theory of Relativity

The Theory of Relativity, developed by Albert Einstein in the early 20th century, represents a monumental shift in our understanding of space, time, and gravity. This theory, composed of the Special Theory of Relativity and the General Theory of Relativity, has provided a foundation for modern physics, shaping our comprehension of the universe and its fundamental laws. In this comprehensive exploration of the Theory of Relativity, we will delve into the historical context, the key principles, and the groundbreaking implications of Einstein's revolutionary ideas.

- **Historical Context: A Brief Overview of Pre-Relativity Physics**

Before delving into the Theory of Relativity, it is essential to understand the state of physics before Einstein's groundbreaking work. In the late 19th century, the predominant view of the physical world was governed by Isaac Newton's laws of motion and James Clerk Maxwell's equations of electromagnetism. While these theories were successful in explaining a wide range of phenomena, they contained inconsistencies and were unable to account for certain experimental results.

One of the most significant challenges faced by physicists of the time was the puzzling behavior of light. Experiments, such as the Michelson-Morley experiment, failed to detect the presence of a luminiferous aether, a hypothetical medium through which light was believed to propagate. This failure suggested that the speed of light might be a universal constant, a radical idea that would become a cornerstone of the Theory of Relativity.

· Special Theory of Relativity: Revolutionizing Space and Time

Albert Einstein published the Special Theory of Relativity in 1905, fundamentally altering our understanding of space and time. This theory emerged from two central postulates:

a) The Principle of Relativity: The laws of physics are the same for all observers in inertial (non-accelerating) reference frames. b) The Constancy of the Speed of Light: The speed of light in a vacuum is the same for all observers, regardless of their motion relative to the light source.

From these postulates, Einstein derived several profound consequences that challenged conventional wisdom:

1. Time Dilation: Time appears to pass more slowly for objects in motion relative to stationary observers. As an object's velocity approaches the speed of light, the effect becomes more pronounced.
2. Length Contraction: Objects in motion contract in the direction of their motion as observed by stationary observers. The faster an object moves, the more significant the contraction.
3. Relativity of Simultaneity: The concept of simultaneity (two events occurring at the same time) is relative and depends on an observer's relative motion. What appears simultaneous to one observer may not be simultaneous to another moving at a different velocity.
4. Mass-Energy Equivalence: The energy of an object is directly proportional to its mass, as expressed by the famous equation $E=mc^2$. This equation implies that mass can be converted into energy and vice versa, as demonstrated by nuclear reactions.
5. General Theory of Relativity: A New Perspective on Gravity

In 1915, Einstein extended the Special Theory of Relativity to include accelerated reference frames, resulting in the General Theory of Relativity. This theory redefined our understanding of gravity, replacing Newton's concept of a gravitational force acting at a distance with the idea that massive objects cause the curvature of spacetime. According to the General Theory of Relativity, objects in freefall follow the shortest path, or geodesic, through this curved spacetime, resulting in the observed effects of gravity.

Key predictions and implications of the General Theory of Relativity include:

Gravitational Time Dilation: Time passes more slowly in stronger gravitational fields. This effect has been confirmed by experiments using atomic clocks placed at different altitudes, with clocks at lower altitudes (closer to Earth's center of mass) running slower than those at higher altitudes.

1. Gravitational Redshift: Light emitted from a massive object, such as a star, is redshifted due to the influence of gravity. This effect, also known as the gravitational redshift of light, causes the light to shift towards longer wavelengths as it escapes the gravitational field.

2. Bending of Light: Light passing through a gravitational field follows the curvature of spacetime, resulting in the bending of light rays. This phenomenon, known as gravitational lensing, has been observed in the deflection of starlight near the Sun and the formation of "Einstein rings" around massive galaxies.

3. Gravitational Waves: Accelerating masses generate ripples in spacetime known as gravitational waves. These waves propagate at the speed of light and can be detected by specialized instruments such as the Laser Interferometer Gravitational-Wave Observatory (LIGO). In 2016, LIGO announced the first direct observation of gravitational waves, providing further confirmation of Einstein's General Theory of Relativity.

4. Black Holes: The General Theory of Relativity predicts the existence of black holes, objects with such immense mass and density that their gravitational pull prevents even light from escaping. Observations of stars orbiting invisible objects and the detection of gravitational waves from merging black holes have provided strong evidence for the

existence of these mysterious cosmic entities.

5. Experimental Verification and the Impact of the Theory of Relativity

Over the past century, numerous experiments and observations have provided strong evidence supporting the predictions of the Theory of Relativity. Some of the most notable examples include:

1. The Eddington Solar Eclipse Expedition (1919): British astronomer Arthur Eddington observed the deflection of starlight near the Sun during a solar eclipse, providing the first experimental confirmation of the General Theory of Relativity.
2. Atomic Clock Experiments: Comparisons of atomic clocks at different altitudes and velocities have verified the effects of time dilation predicted by both the Special and General Theories of Relativity.
3. Global Positioning System (GPS): The accurate functioning of the GPS network relies on accounting for the effects of both special and general relativistic time dilation on satellite clocks.
4. The Hafele-Keating Experiment (1971): This experiment involved flying atomic clocks around the world on commercial airliners and comparing their timekeeping with clocks on the ground, providing further evidence of time dilation due to both relative motion and gravitational effects.
5. The Impact on Modern Physics: The Theory of Relativity has had a profound impact on various fields of physics, including cosmology, astrophysics, and particle physics. It has also played a crucial role in shaping our understanding of the universe and its fundamental laws, paving the way for the development of the Standard Model of particle physics and the current quest for a unified theory of all fundamental forces.

The Theory of Relativity, developed by Albert Einstein, represents a groundbreaking shift in our understanding of the fundamental nature of space, time, and gravity. This theory has provided the foundation for modern physics, shaping our comprehension of the universe and influencing count-less technological advances. Through its far-reaching implications, the Theory of Relativity has not only revolutionized the field of physics but also fundamentally altered the way we perceive the cosmos and our place within it.

As we continue to explore the mysteries of the universe, the Theory of Relativity will undoubtedly remain a cornerstone of our scientific endeavors. It serves as a testament to human curiosity and our relentless pursuit of knowledge, inspiring future generations of physicists and thinkers to probe the depths of reality and unravel the secrets of the cosmos.

9

Time Dilation and Length Contraction

Two of the most fascinating consequences of the Special Theory of Relativity, developed by Albert Einstein in 1905, are time dilation and length contraction. These phenomena challenge our intuitive understanding of time and space, demonstrating that they are not absolute but relative to an observer's motion. In this comprehensive analysis, we will explore the theoretical foundations, experimental evidence, and real-world implications of time dilation and length contraction.

· **The Theoretical Basis of Time Dilation and Length Contraction**

Time dilation and length contraction arise from the two central postulates of the Special Theory of Relativity:

a) The Principle of Relativity: The laws of physics are the same for all observers in inertial (non-accelerating) reference frames. b) The Constancy of the Speed of Light: The speed of light in a vacuum is the same for all observers, regardless of their motion relative to the light source.

1.1 Time Dilation

Time dilation is the phenomenon in which time appears to pass more slowly for objects in motion relative to stationary observers. This effect becomes more pronounced as an object's velocity approaches the speed of light. Mathematically, time dilation can be expressed using the Lorentz factor

(γ):

$\gamma = 1 / \sqrt{(1 - v^2/c^2)}$

Here, v represents the relative velocity between the moving object and the stationary observer, and c is the speed of light. The time interval experienced by the moving object (t') is related to the time interval measured by the stationary observer (t) by the following equation:

$t' = t / \gamma$

As the relative velocity (v) approaches the speed of light (c), the Lorentz factor (γ) approaches infinity, causing the moving object's time to appear infinitely dilated (slowed down) compared to the stationary observer's time.

1.2 Length Contraction

Length contraction is the phenomenon in which objects in motion contract in the direction of their motion as observed by stationary observers. The faster an object moves, the more significant the contraction. Length contraction can also be expressed using the Lorentz factor (γ):

$L' = L / \gamma$

In this equation, L represents the length of the object as measured by a stationary observer, while L' denotes the length of the object as experienced by an observer moving along with it. As the relative velocity (v) approaches the speed of light (c), the Lorentz factor (γ) approaches infinity, causing the moving object's length to appear contracted to zero in the direction of motion.

2. Experimental Evidence for Time Dilation and Length Contraction

Numerous experiments have been conducted to test the predictions of time dilation and length contraction, providing strong evidence in support of these relativistic effects.

2.1 Time Dilation Experiments

One of the most famous time dilation experiments is the 1971 Hafele-Keating experiment. Atomic clocks were flown around the world on commercial airliners, and their timekeeping was compared to atomic clocks on the ground. The results confirmed that the moving clocks experienced time dilation, running slower than their stationary counterparts, in agreement with the Special Theory of Relativity.

Another significant experiment involving time dilation is the observation of muons. Muons are subatomic particles with a relatively short half-life. When cosmic rays strike Earth's atmosphere, they produce muons, which travel at near-light speeds toward the surface. Due to time dilation, muons experience a longer half-life from the perspective of a stationary observer on Earth, allowing more of them to reach the surface than would be expected without relativistic effects.

2.2 Length Contraction Experiments

Length contraction has proven more challenging to measure directly due to the minuscule nature of the effect at everyday velocities. However, indirect evidence for length contraction can be found in experiments involving particle accelerators. In these experiments, charged particles, such as electrons and protons, are accelerated to velocities approaching the speed of light. The magnetic fields used to guide these particles take into account the predicted length contraction, ensuring that the particles remain on their intended paths. The success of these experiments indirectly confirms the presence of length contraction.

Additionally, high-energy collisions in particle accelerators produce particles traveling at relativistic speeds. These particles are detected using specialized detectors designed based on the predicted effects of length contraction. The accurate detection and analysis of such particles provide further indirect evidence for the existence of length contraction.

3.Real-World Implications and Applications of Time Dilation and Length Contraction

While the effects of time dilation and length contraction may seem confined to the realm of theoretical physics or extreme conditions, they have real-world implications and applications in various fields.

3.1 Global Positioning System (GPS)

The Global Positioning System (GPS) relies on a network of satellites orbiting Earth to provide precise location information. These satellites move at high velocities relative to observers on the ground, and they also experience different gravitational fields due to their altitude. Both time dilation and length contraction must be accounted for in the satellite clocks and signal processing to ensure the accurate functioning of the GPS.

3.2 Particle Accelerators and High-Energy Physics

As previously mentioned, particle accelerators rely on the principles of time dilation and length contraction to guide and analyze high-energy particles. These accelerators have been instrumental in the discovery of fundamental particles and the development of the Standard Model of particle physics.

3.3 Space Travel and Interstellar Voyages

For hypothetical interstellar voyages involving speeds close to the speed of light, time dilation would play a significant role. Travelers aboard a spacecraft moving at relativistic speeds would experience time dilation, aging more slowly compared to people remaining on Earth. This effect, often referred to as the "twin paradox," raises intriguing questions about the nature of time and the practicality of long-distance space travel.

Time dilation and length contraction, two remarkable consequences of the Special Theory of Relativity, have reshaped our understanding of space and time. These relativistic effects, confirmed by numerous experiments, demonstrate that our perception of time and space is not absolute but depends on our motion relative to other objects. As we continue to push the boundaries of our knowledge and technology, the principles of time dilation and length contraction will undoubtedly play a crucial role in shaping our understanding of the universe and our place within it.

10

The Geometry of Space-Time

The geometry of space-time is a fundamental concept in the study of the universe, serving as a cornerstone for our understanding of relativity, gravity, and cosmology. Developed from the groundbreaking work of Albert Einstein and other physicists, this concept combines space and time into a single four-dimensional continuum, allowing for a unified description of the behavior of matter and energy. In this comprehensive examination, we will delve into the origins, mathematical framework, and profound implications of the geometry of space-time.

1.The Origins of Space-Time Geometry: From Newtonian Mechanics to the Theory of Relativity

The development of space-time geometry has its roots in the evolution of our understanding of space, time, and gravity, beginning with the classical Newtonian mechanics and culminating in Einstein's Theory of Relativity.

1.1 Newtonian Mechanics and Absolute Space and Time

In the late 17th century, Isaac Newton formulated the laws of motion and the universal law of gravitation, which provided a comprehensive description of the motion of celestial and terrestrial objects. Central to Newton's mechanics were the notions of absolute space and time, which were considered to be independent and immutable.

1.2 The Emergence of Electromagnetism and the Aether Hypothesis

By the late 19th century, the discovery of electromagnetism and the formulation of Maxwell's equations led to a new understanding of the behavior of electric and magnetic fields. However, these developments also introduced inconsistencies with the Newtonian framework. In particular, the assumption of an absolute space and time was incompatible with the observation that the speed of light appeared constant for all observers, regardless of their relative motion.

To reconcile this inconsistency, physicists proposed the existence of a luminiferous aether, a hypothetical medium through which light was thought to propagate. However, the famous Michelson-Morley experiment failed to detect any evidence of the aether, suggesting that the speed of light was indeed a universal constant.

1.3 The Special Theory of Relativity and the Birth of Space-Time

In 1905, Albert Einstein introduced the Special Theory of Relativity, which resolved the inconsistencies between classical mechanics and electromagnetism. The theory was based on two postulates:

a) The Principle of Relativity: The laws of physics are the same for all observers in inertial (non-accelerating) reference frames. b) The Constancy of the Speed of Light: The speed of light in a vacuum is the same for all observers, regardless of their motion relative to the light source.

Einstein's theory unified space and time into a single four-dimensional continuum known as space-time. In this framework, space and time were no longer absolute but were relative to the motion of observers. The Special Theory of Relativity also revealed that the geometry of space-time was flat or Minkowskian, characterized by a four-dimensional analog of Euclidean geometry.

1.4 The General Theory of Relativity and Curved Space-Time

In 1915, Einstein expanded upon the Special Theory of Relativity to incorporate gravity, giving rise to the General Theory of Relativity. This theory replaced the Newtonian concept of gravity as a force acting at a distance with the idea that massive objects cause the curvature of space-time. Objects moving in a gravitational field follow the shortest path or geodesic through this curved space-time, resulting in the observed effects of gravity. The General Theory of Relativity introduced the concept of curved space-time, which is described by a mathematical framework known as Riemannian geometry.

2.The Mathematical Framework of Space-Time Geometry

The geometry of space-time is a crucial aspect of both the Special and General Theories of Relativity, providing the foundation for describing the behavior of matter and energy in the universe. The mathematical framework for space-time geometry involves several key elements, including metrics, tensors, and differential geometry.

2.1 Metrics and the Space-Time Interval

The metric is a fundamental component of space-time geometry, describing the distance between events in space-time. In the Special Theory of Relativity, the Minkowski metric is used, which is based on a flat, four-dimensional space-time. The Minkowski metric defines the space-time interval (s^2) between two events using the following equation:

$$s^2 = c^2 \Delta t^2 - \Delta x^2 - \Delta y^2 - \Delta z^2$$

Here, Δt, Δx, Δy, and Δz represent the differences in time and spatial coordinates between the events, and c is the speed of light. The space-time interval is invariant, meaning it remains constant for all observers regardless of their relative motion.

In the General Theory of Relativity, the metric is more complex, as it must account for the curvature of space-time due to the presence of mass and energy. The metric in curved space-time is given by the metric tensor, which is a symmetric 4x4 matrix that describes the geometry of space-time at each point.

2.2 Tensors and the Stress-Energy-Momentum Tensor

Tensors are mathematical objects used to describe the properties of space-time and the distribution of mass and energy within it. They have the unique property of being invariant under coordinate transformations, making them particularly suited for describing the relativistic behavior of objects.

In the General Theory of Relativity, the stress-energy-momentum tensor ($T\mu\nu$) is a crucial element, representing the distribution of mass, energy, and momentum in space-time. This tensor is a 4x4 matrix that includes components such as energy density, momentum density, and stress (pressure and shear forces).

2.3 Differential Geometry and the Einstein Field Equations

Differential geometry is a branch of mathematics that deals with the study of curved spaces and the properties of objects within them. It is an essential tool for understanding the geometry of space-time in the context of the General Theory of Relativity.

The cornerstone of the General Theory of Relativity is the Einstein field equations, a set of ten partial differential equations that relate the metric tensor (which describes the curvature of space-time) to the stress-energy-momentum tensor (which describes the distribution of mass and energy). These equations govern the behavior of space-time and the motion of objects within it, providing a comprehensive description of gravitational phenomena.

3. Implications and Applications of Space-Time Geometry

The geometry of space-time has far-reaching implications for our understanding of the universe, from the smallest subatomic particles to the largest cosmic structures. It also plays a crucial role in a variety of practical applications, including the Global Positioning System (GPS) and the study of black holes and cosmology.

3.1 Black Holes and Event Horizons

Black holes are astronomical objects with such immense gravitational pull that nothing, not even light, can escape their grasp. The General Theory of Relativity predicts the existence of black holes as solutions to the Einstein field equations, revealing that they are regions of space-time where the curvature becomes extreme.

The event horizon is a critical concept in the study of black holes, representing the boundary within which the gravitational pull is so strong that escape is impossible. The geometry of space-time near the event horizon is characterized by a significant warping, leading to fascinating phenomena such as gravitational time dilation and the stretching of objects, known as "spaghettification."

3.2 Cosmology and the Expanding Universe

The geometry of space-time also plays a central role in our understanding of the large-scale structure and evolution of the universe. The Friedmann-Lemaître-R obertson-Walker (FLRW) metric is a solution to the Einstein field equations that describes an isotropic and homogeneous universe. The FLRW metric includes a scale factor that characterizes the expansion or contraction of the universe, which is governed by the distribution of mass and energy within it.

Observations of distant supernovae and the cosmic microwave background radiation have provided strong evidence for an expanding universe, with the current consensus being that the universe is undergoing accelerated expansion due to a mysterious form of energy known as dark energy.

3.3 Gravitational Waves and the Detection of Cosmic Events

Gravitational waves are ripples in the fabric of space-time, generated by the acceleration of massive objects such as merging black holes or neutron stars. The existence of gravitational waves was predicted by the General Theory of Relativity, and their detection has become a major focus of experimental physics in recent years.

In 2016, the Laser Interferometer Gravitational-Wave Observatory (LIGO) made the groundbreaking detection of gravitational waves, confirming a key prediction of the General Theory of Relativity and opening a new window into the study of the universe. Gravitational wave detectors such as LIGO and its counterparts around the world are now providing valuable insights into the geometry of space-time and the properties of cosmic events that were previously inaccessible through electromagnetic observations alone.

4.Challenges and Future Perspectives in the Study of Space-Time Geometry

The geometry of space-time has revolutionized our understanding of the universe, providing a consistent framework for describing the behavior of matter, energy, and gravity. However, several challenges and open questions remain, driving ongoing research and exploration.

4.1 Quantum Gravity and the Unification of Physics

One of the most significant challenges in modern physics is the development of a quantum theory of gravity that unifies the General Theory of Relativity with quantum mechanics. While both theories have been incredibly successful in describing their respective domains (gravity on large scales and the behavior of particles on small scales), they are fundamentally incompatible, leading to inconsistencies and infinities in certain calculations.

The geometry of space-time is central to this quest for unification, with various approaches such as string theory, loop quantum gravity, and emergent gravity seeking to reconcile the two theories and provide a consistent description of the universe at all scales.

4.2 Dark Energy and the Fate of the Universe

The discovery of the accelerated expansion of the universe has raised profound questions about the nature of dark energy and the ultimate fate of the cosmos. The geometry of space-time is intimately connected to these questions, with the distribution of mass and energy determining the large-scale structure and evolution of the universe.

Understanding the nature of dark energy and its impact on the geometry of space-time is a major focus of contemporary cosmology, with ongoing observations and experiments aimed at probing the properties of this mysterious form of energy and its implications for the future of the universe.

The geometry of space-time has fundamentally reshaped our understanding of the universe, unifying space and time into a single continuum and providing a comprehensive framework for describing the behavior of matter, energy, and gravity. From the smallest subatomic particles to the largest cosmic structures, the geometry of space-time continues to play a crucial role in our quest to unravel the mysteries of the cosmos and to develop a complete and consistent description of the physical world. As we forge ahead in our exploration of the universe, the study of space-time geometry remains an essential and fascinating endeavor, holding the promise of new insights and breakthroughs in our understanding of the fabric of reality.

III

The Expanding Universe

11

Introduction to the Expanding Universe

The concept of an expanding universe is a fundamental aspect of modern cosmology, providing the foundation for our understanding of the large-scale structure and evolution of the cosmos. This groundbreaking idea, rooted in the General Theory of Relativity and supported by a wealth of observational evidence, has revolutionized our perception of the universe and its history. In this extensive examination, we will explore the theoretical and observational aspects of the expanding universe, delving into its origins, the evidence supporting it, its implications for cosmic evolution, and the ongoing challenges and questions that drive the field of cosmology.

1. **The Origins of the Expanding Universe: From Static Models to Cosmic Expansion**

The notion of an expanding universe emerged from the interplay between theoretical predictions and observational discoveries, challenging the long-held view of a static, unchanging cosmos.

1.1 The Einstein Static Universe and the Cosmological Constant

In the early 20th century, Albert Einstein developed the General Theory of Relativity, a revolutionary new framework for understanding gravity as the curvature of space-time. In an attempt to apply his theory to the universe as a whole, Einstein proposed a static model of the cosmos, in which the universe had no overall expansion or contraction.

However, Einstein's original equations did not support a static universe due to the gravitational attraction between its constituent masses. To resolve this issue, he introduced a term known as the cosmological constant (Λ) into his equations, which represented a form of repulsive force that counterbalanced the attractive force of gravity. This allowed for the construction of a static model of the universe, known as the Einstein static universe.

1.2 The Emergence of the Expanding Universe: Friedmann, Lemaître, and Hubble

Despite Einstein's efforts to create a static model of the universe, the notion of cosmic expansion soon gained traction due to the work of several key figures.

In the 1920s, Russian mathematician Alexander Friedmann and Belgian physicist Georges Lemaître independently derived solutions to Einstein's equations that predicted an expanding universe. These solutions, now known as the Friedmann-Lemaître-Robertson-Walker (FLRW) metric, described a universe in which the overall scale factor (a measure of the size of the universe) varied with time. This was a radical departure from the static models of the time and laid the groundwork for the concept of an expanding universe.

In 1929, American astronomer Edwin Hubble provided the first observational evidence for cosmic expansion. By studying the spectra of distant galaxies, Hubble discovered a systematic redshift in their light, which indicated that they were moving away from us. Moreover, he found that the recessional velocity of these galaxies was proportional to their distance, a relationship now known as Hubble's law. Hubble's observations lent strong support to the notion of an expanding universe, prompting a fundamental shift in our understanding of the cosmos.

2.Observational Evidence for the Expanding Universe

In the decades since Hubble's discovery, a wealth of observational evidence has been amassed, which has firmly established the concept of an expanding universe.

2.1 The Cosmic Microwave Background Radiation

The cosmic microwave background (CMB) radiation is the relic radiation from the early universe, dating back to a time when the universe was dense, hot, and filled with a plasma of charged particles and photons. As the universe expanded and cooled, the plasma eventually became neutral atoms, allowing the photons to travel freely through space.

The discovery of the CMB in 1964 by Arno Penzias and Robert Wilson provided strong evidence for the Big Bang theory, which posits that the universe began in a hot, dense state and has been expanding and cooling ever since. The uniformity and near-perfect blackbody spectrum of the CMB are consistent with the predictions of an expanding universe, and its properties provide crucial insights into the early history and evolution of the cosmos.

2.2 The Large-Scale Structure of the Universe

The large-scale structure of the universe, composed of galaxies, galaxy clusters, and cosmic filaments, provides further evidence for an expanding universe. Observations of the distribution of galaxies and the patterns of their motion are consistent with the predictions of the FLRW metric, indicating that the universe is expanding and evolving over time.

Moreover, the clustering of galaxies and the cosmic web of large-scale structures are consistent with the theoretical predictions of cosmic expansion and the growth of structures through gravitational instability. These observations support the idea that cosmic expansion has played a fundamental role in shaping the large-scale structure of the universe.

2.3 Supernovae and the Accelerating Expansion

Observations of distant Type Ia supernovae have provided compelling evidence not only for the expansion of the universe but also for its acceleration. In the late 1990s, two independent research teams discovered that distant Type Ia supernovae were dimmer than expected, implying that they were farther away than predicted by a uniformly expanding universe.

This surprising observation suggested that the expansion of the universe is accelerating, driven by an unknown form of energy referred to as dark energy. The discovery of the accelerating expansion of the universe has profound implications for our understanding of the cosmos, and the nature of dark energy remains one of the greatest mysteries in modern physics.

3.The Evolution of the Universe: From the Big Bang to the Present

The concept of an expanding universe provides the foundation for our understanding of the history and evolution of the cosmos, from the moment of the Big Bang to the present day.

3.1 The Early Universe and the Hot Big Bang

According to the Big Bang theory, the universe began in a hot, dense state approximately 13.8 billion years ago. As the universe expanded, it cooled, allowing for the formation of subatomic particles, followed by the synthesis of light elements such as hydrogen, helium, and lithium during a period known as Big Bang nucleosynthesis.

Approximately 380,000 years after the Big Bang, the universe became transparent to radiation as the charged plasma combined to form neutral atoms. This event marked the last scattering of photons, which we now observe as the cosmic microwave background radiation.

3.2 The Formation of Galaxies and Large-Scale Structures

As the universe continued to expand and cool, the slight density fluctuations present in the early universe grew through gravitational instability, eventually leading to the formation of the first stars and galaxies. The hierarchical process of structure formation, driven by the interplay between gravity and cosmic expansion, shaped the complex cosmic web of galaxies, clusters, and filaments that we observe today.

3.3 The Accelerating Universe and the Role of Dark Energy

The discovery of the accelerating expansion of the universe has raised fascinating questions about the nature of dark energy and its role in cosmic evolution. While the precise nature of dark energy remains unknown, its presence has profound implications for the future of the universe, as it drives an ever-accelerating expansion that will ultimately determine the fate of the cosmos.

4.Challenges and Future Prospects in the Study of the Expanding Universe

Despite the wealth of evidence supporting the expanding universe and the progress made in understanding its evolution, several challenges and open questions remain, spurring ongoing research and exploration in the field of cosmology.

4.1 The Nature of Dark Energy

Understanding the nature of dark energy is one of the most significant challenges in modern cosmology. Various theoretical models have been proposed to explain dark energy, ranging from a cosmological constant to dynamical scalar fields to modifications of General Relativity. Determining the precise nature of dark energy and its impact on the expansion of the universe is a major focus of contemporary research, with ongoing observations and experiments aimed at probing its properties and shedding light on this enigmatic form of energy.

4.2 The Initial Conditions and Inflation

The initial conditions of the universe and the mechanisms that generated the primordial density fluctuations responsible for the formation of large-scale structures are still not fully understood. The theory of cosmic inflation, which posits that the universe underwent a rapid, exponential expansion in the first fraction of a second after the Big Bang, has been proposed as a possible solution to several puzzles in the standard Big Bang model, including the origin of these density fluctuations.

While inflation is supported by several lines of evidence, such as the near-uniformity of the CMB and the observed pattern of anisotropies, the details of the inflationary process and the underlying physics are still the subject of ongoing research and debate.

4.3 The Integration of Dark Matter and Baryonic Matter

The role of dark matter, a mysterious form of matter that interacts primarily through gravity and makes up approximately 27% of the universe's mass-energy content, is another key aspect of the expanding universe. The interplay between dark matter and ordinary baryonic matter plays a crucial role in the formation and evolution of cosmic structures, and understanding the nature of dark matter and its impact on the expansion of the universe is a significant challenge in modern cosmology.

4.4 The Quest for a Quantum Theory of Gravity

The development of a quantum theory of gravity that unifies General Relativity with quantum mechanics is an important goal in the pursuit of a complete and consistent description of the universe. While both theories have been incredibly successful in their respective domains, they are fundamentally incompatible, leading to inconsistencies and infinities in certain calculations. A quantum theory of gravity would have profound implications for our understanding of the early universe, black holes, and the ultimate fate of the cosmos.

The concept of an expanding universe is a cornerstone of modern cosmology, providing the foundation for our understanding of the large-scale structure and evolution of the cosmos. From its origins in the work of Friedmann, Lemaître, and Hubble, to the wealth of observational evidence that has been amassed in the decades since, the expanding universe has revolutionized our perception of the cosmos and its history.

As we continue to explore the universe and probe its mysteries, the study of cosmic expansion remains a vibrant and dynamic field, offering the promise of new insights and breakthroughs in our understanding of the fundamental nature of the universe and its evolution.

12

Edwin Hubble and the Redshift

Edwin Hubble, an American astronomer, played a transformative role in the field of astronomy and cosmology through his groundbreaking discovery of the redshift of light from distant galaxies. This observation, which demonstrated that galaxies are moving away from us, laid the foundation for the concept of an expanding universe and altered our understanding of the cosmos. In this comprehensive exploration, we will delve into the life and work of Edwin Hubble, the science behind redshift, and the far-reaching implications of his seminal discovery.

1. Edwin Hubble: A Life in Astronomy

Edwin Hubble's journey in astronomy began with an innate curiosity about the cosmos and culminated in a series of groundbreaking discoveries that forever changed our understanding of the universe.

1.1 Early Life and Education

Edwin Powell Hubble was born on November 20, 1889, in Marshfield, Missouri, and spent his early years in Illinois. As a young boy, he was fascinated by the stars and often spent hours gazing at the night sky through a telescope given to him by his grandfather.

Hubble attended the University of Chicago, where he studied mathematics and astronomy. In 1910, he was awarded a Rhodes Scholarship to study at the

University of Oxford in England, where he pursued a degree in jurisprudence. Although his academic focus shifted away from astronomy during his time at Oxford, his passion for the subject remained undiminished.

1.2 A Career in Astronomy

After serving in World War I, Hubble returned to the United States and resumed his studies in astronomy. In 1919, he earned a Ph.D. from the University of Chicago, with a dissertation on the distribution of galaxies in space. Following the completion of his doctorate, Hubble was offered a position at the Mount Wilson Observatory in California, which housed the world's largest telescope at the time.

It was at Mount Wilson that Hubble made his most significant contributions to astronomy, investigating the nature of galaxies and their motion through space. His work laid the groundwork for the modern field of extragalactic astronomy and the study of cosmic evolution.

2The Science of Redshift and the Expanding Universe

To fully appreciate Hubble's discovery, it is essential to understand the underlying principles of redshift and the Doppler effect, which form the basis for the observation of cosmic expansion.

2.1 The Doppler Effect and Redshift

The Doppler effect is a phenomenon that occurs when the frequency of a wave, such as sound or light, changes due to the relative motion between the source of the wave and the observer. When the source of a wave is moving away from the observer, the observed frequency decreases, and the wavelengths become longer. Conversely, when the source is moving towards the observer, the frequency increases, and the wavelengths become shorter.

In the context of light waves, the Doppler effect causes a shift in the observed color of the light. When the source of light is moving away from the observer, the wavelengths become longer, and the light appears more red than its original color. This phenomenon is known as redshift. Conversely, when the source is moving towards the observer, the wavelengths become shorter, and the light appears more blue than its original color. This phenomenon is known as blueshift.

2.2 Spectroscopy and the Measurement of Redshift

Spectroscopy is the study of the interaction of light with matter, allowing for the analysis of the properties of celestial objects by examining their emitted or absorbed light. One of the key techniques in spectroscopy is the analysis of absorption or emission lines, which are specific wavelengths of light that correspond to the energy levels of atoms or molecules in the object.

By comparing the observed wavelengths of these spectral lines to their known laboratory values, astronomers can measure the redshift or blueshift of the light and determine the relative motion of the celestial object. This method is particularly useful for studying the motion of distant galaxies, as their light contains numerous characteristic spectral lines that can be used to measure their redshift.

3.Hubble's Discovery: The Redshift-Distance Relationship

Working at the Mount Wilson Observatory, Edwin Hubble made a series of observations that led to the discovery of the redshift-distance relationship, which would ultimately reshape our understanding of the universe.

3.1 The Classification and Distances of Galaxies

One of Hubble's first significant contributions to astronomy was the classification of galaxies according to their appearance. In the 1920s, he developed a system known as the Hubble Sequence, which organized galaxies into categories based on their shape and structure, such as ellipticals, spirals, and barred spirals.

In addition to classifying galaxies, Hubble was also interested in determining their distances. He employed the method of Cepheid variable stars, first established by Henrietta Swan Leavitt, as a means of measuring the distances to nearby galaxies. Cepheid variables are a type of pulsating star with a well-defined relationship between their brightness and pulsation period, which allows astronomers to use them as standard candles to estimate distances.

3.2 The Discovery of the Redshift-Distance Relationship

In 1929, Hubble made the groundbreaking observation that would forever change our view of the cosmos. By measuring the redshift of light from distant galaxies and comparing it to their estimated distances, he discovered that there was a linear relationship between the two. This relationship, now known as Hubble's Law, states that the recessional velocity of a galaxy is proportional to its distance from us.

Hubble's Law implies that galaxies are moving away from us, and the farther away a galaxy is, the faster it is receding. This observation provided the first compelling evidence for the concept of an expanding universe, in which the overall scale of the universe is increasing with time.

4.The Impact and Legacy of Hubble's Discovery

Hubble's discovery of the redshift-distance relationship and its implications for an expanding universe have had far-reaching consequences for the fields of astronomy and cosmology.

4.1 The Expansion of the Universe

The notion of an expanding universe, which emerged from Hubble's observations, has become a cornerstone of modern cosmology. It provides the foundation for our understanding of the large-scale structure and evolution of the cosmos, from the moment of the Big Bang to the present day.

The expanding universe also has profound implications for the future of the cosmos, as the nature of cosmic expansion determines the ultimate fate of the universe. The discovery of the accelerating expansion of the universe in the late 1990s, driven by the mysterious dark energy, has raised fascinating questions about the destiny of the cosmos and the nature of the forces that govern it.

4.2 The Birth of Extragalactic Astronomy

Hubble's work on the classification and distances of galaxies, along with his discovery of the redshift-distance relationship, laid the groundwork for the field of extragalactic astronomy. The study of galaxies beyond the Milky Way has since become a vibrant and dynamic area of research, providing valuable insights into the formation, evolution, and clustering of galaxies in the universe.

Edwin Hubble's discovery of the redshift of light from distant galaxies and its relationship to their distances represents a monumental breakthrough in the history of astronomy and cosmology. This observation, which demonstrated that the universe is expanding, has reshaped our understanding of the cosmos and its evolution, providing a foundation for the modern field of extragalactic astronomy and the study of cosmic expansion.

13

The Big Bang Theory

The Big Bang Theory is the prevailing cosmological model that describes the origin and evolution of the universe. It posits that the universe began as an extremely hot and dense state and has since expanded and cooled over the course of approximately 13.8 billion years. The Big Bang Theory has been refined and supported by a wealth of observational evidence, making it the most widely accepted explanation for the origin of the cosmos. In this comprehensive exploration, we will examine the development of the Big Bang Theory, the key evidence supporting it, and the implications it has for our understanding of the universe.

1.Historical Background and Development of the Big Bang Theory

The Big Bang Theory has its roots in the early 20th century and has evolved significantly over the past century as new observations and theoretical developments have shaped our understanding of the cosmos.

1.1 Early Theoretical Foundations

The seeds of the Big Bang Theory can be traced back to the early 20th century with the development of Albert Einstein's General Theory of Relativity, which provides a framework for describing the gravitational force and the geometry of spacetime. In 1922, Russian mathematician Alexander Friedmann derived a set of equations from Einstein's theory that described an expanding universe. Around the same time, Belgian astronomer Georges

Lemaître independently developed a similar model, which he referred to as the "hypothesis of the primeval atom."

1.2 Observational Evidence for an Expanding Universe

The idea of an expanding universe gained traction in the late 1920s with the work of American astronomer Edwin Hubble. Through his observation of the redshift of light from distant galaxies, Hubble discovered a relationship between the distance of a galaxy and its recessional velocity. This relationship, now known as Hubble's Law, provided compelling evidence that the universe is expanding, lending support to the earlier theoretical work of Friedmann and Lemaître.

1.3 The Emergence of the Big Bang Theory

In the decades following Hubble's discovery, the concept of an expanding universe was further developed and refined by numerous scientists, including George Gamow, Ralph Alpher, and Robert Herman. They proposed that the universe began as a hot, dense state and expanded and cooled over time, eventually allowing for the formation of atoms, stars, and galaxies.

The term "Big Bang" was coined by British astronomer Fred Hoyle during a 1949 radio broadcast. Initially intended as a pejorative term to criticize the theory, it was eventually adopted by scientists and popular culture to describe the model.

2. Key Evidence Supporting the Big Bang Theory

The Big Bang Theory has been corroborated by a wealth of observational evidence, which has strengthened its position as the dominant cosmological model.

2.1 Cosmic Microwave Background Radiation

One of the most significant pieces of evidence for the Big Bang Theory is the Cosmic Microwave Background (CMB) radiation. The CMB is the afterglow of the early universe, a nearly uniform radiation field that permeates all of space. It was first detected in 1964 by American astronomers Arno Penzias and Robert Wilson, who were awarded the Nobel Prize in Physics for their discovery.

The CMB is consistent with the predictions of the Big Bang Theory, which posits that the early universe was filled with a hot, dense plasma that emitted radiation. As the universe expanded and cooled, this radiation was stretched and redshifted, eventually becoming the microwave radiation that we observe today.

2.2 Nucleosynthesis and the Abundance of Light Elements

Another key piece of evidence for the Big Bang Theory is the observed abundance of light elements, such as hydrogen, helium, and lithium. The theory predicts that these elements were formed during aperiod known as Big Bang nucleosynthesis, which occurred when the universe was still very young and hot. As the universe cooled, protons and neutrons combined to form the nuclei of these light elements in specific proportions that depend on the conditions of the early universe.

The observed abundances of these elements in the universe are in remarkable agreement with the predictions of Big Bang nucleosynthesis, providing strong support for the theory. In particular, the observed ratio of helium to hydrogen, which is approximately 1:12 by mass, is consistent with the expectations of the Big Bang model.

2.3 The Large-Scale Structure of the Universe

The distribution and evolution of large-scale structures in the universe, such as galaxies and galaxy clusters, provide further evidence for the Big Bang Theory. The theory predicts that the initial conditions of the universe contained tiny fluctuations in density, which grew over time due to gravitational attraction. These density fluctuations eventually led to the formation of the large-scale structures that we observe today.

Observations of the distribution of galaxies and their motion, as well as the distribution of mass in the universe, support the predictions of the Big Bang model. Additionally, computer simulations of cosmic structure formation based on the principles of the Big Bang Theory have been successful in reproducing the observed properties of the universe.

3.The Early Universe and Cosmic Inflation

The Big Bang Theory has been refined and extended to include a description of the very early universe, during which a rapid, exponential expansion known as cosmic inflation is believed to have occurred.

3.1 The Need for Inflation

Cosmic inflation was proposed as a solution to several puzzles in the standard Big Bang model, such as the horizon problem and the flatness problem. The horizon problem arises from the observation that the CMB has a nearly uniform temperature across the sky, despite the fact that regions separated by large distances would not have had time to exchange energy and reach thermal equilibrium in the standard model. The flatness problem concerns the observation that the geometry of the universe is very close to being flat, which requires a fine-tuning of the initial conditions in the standard model.

3.2 The Inflationary Model

In 1980, American physicist Alan Guth proposed the theory of cosmic inflation, which posits that the universe underwent a brief period of rapid, exponential expansion in the first fraction of a second after the Big Bang. This rapid expansion would have stretched and smoothed out the initial density fluctuations, explaining the observed uniformity of the CMB and the flatness of the universe.

The theory of inflation has been further refined by other physicists, such as Andrei Linde and Paul Steinhardt, and is now an integral part of the modern understanding of the early universe. Inflation also provides a mechanism for generating the primordial density fluctuations that seeded the formation of large-scale structures in the universe.

4. The Future of the Universe and Open Questions

The Big Bang Theory, while successful in explaining many aspects of the universe, also raises new questions and challenges that are the subject of ongoing research and debate.

4.1 The Accelerating Expansion of the Universe

In the late 1990s, observations of distant supernovae provided evidence that the expansion of the universe is accelerating, a discovery that was awarded the Nobel Prize in Physics in 2011. This accelerating expansion is believed to be driven by a mysterious form of energy known as dark energy, which makes up approximately 68% of the universe's mass-energy content. The nature of dark energy and its implications for the future of the universe remain open questions in modern cosmology.

4.2 The Nature of Dark Matter

Another challenge in the Big Bang Theory is the nature of dark matter, which accounts for about 27% of the universe's mass-energy content. Dark matter is a form of matter that interacts weakly with electromagnetic radiation, making it difficult to detect directly. Its presence is inferred from its gravitational effects on the motion of galaxies and the large-scale structure of the universe. The composition and properties of dark matter are still unknown, and its identification remains a key goal in modern astrophysics and particle physics.

4.3 The Initial Conditions of the Universe

One of the major unsolved questions in cosmology concerns the initial conditions of the universe. The Big Bang Theory describes the evolution of the universe from a hot, dense state but does not explain how the universe came into existence or what preceded the Big Bang. Some theories propose that the universe emerged from a quantum fluctuation or that it is part of a larger multiverse, but these ideas remain speculative and unproven.

The Big Bang Theory has emerged as the leading cosmological model for the origin and evolution of the universe, supported by a wealth of observational evidence and theoretical developments. From the discovery of the expanding universe to the detection of the cosmic microwave background radiation, the Big Bang Theory has provided a coherent and successful framework for understanding the history and structure of the cosmos.

Despite its many successes, the Big Bang Theory also raises new questions and challenges, such as the nature of dark matter and dark energy, the initial conditions of the universe, and the possible existence of a multiverse. As our understanding of the cosmos continues to advance, the Big Bang Theory will undoubtedly be refined and extended, providing ever deeper insights into the mysteries of the universe and its origins.

14

Cosmic Microwave Background Radiation

The Cosmic Microwave Background (CMB) radiation is a critical piece of observational evidence supporting the Big Bang Theory, which posits that the universe began in a hot, dense state and has since expanded and cooled. The CMB is the residual radiation from the early universe, a nearly uniform field of microwave radiation that permeates all of space. Its discovery, properties, and implications for our understanding of the cosmos have made the CMB a central focus of modern cosmology. In this comprehensive exploration, we will delve into the origins of the CMB, its discovery, and the insights it has provided into the universe's history and structure.

1. **Origins of the Cosmic Microwave Background Radiation**

The Cosmic Microwave Background radiation is an afterglow of the early universe when it was still hot and dense, filled with a plasma of photons, electrons, and atomic nuclei.

1.1 Recombination and the Release of Photons

Around 380,000 years after the Big Bang, the universe had expanded and cooled enough for electrons and atomic nuclei to combine and form neutral atoms, primarily hydrogen and helium. This process, known as recombination, marked a significant transition in the history of the universe.

Before recombination, photons were scattered frequently by the free electrons in the plasma, preventing them from traveling very far. When the electrons combined with nuclei to form neutral atoms, the photons were suddenly free to travel unimpeded across the universe. These photons constitute the CMB radiation we observe today.

1.2 Redshifting and the Cooling of the CMB

As the universe continued to expand, the CMB photons were redshifted, causing their wavelengths to stretch and their energy to decrease. This redshifting has caused the CMB's temperature to drop over time. Today, the CMB has a nearly uniform temperature of approximately 2.725 Kelvin, making it a cold relic of the universe's hot and dense beginnings.

2.The Discovery of the Cosmic Microwave Background Radiation

The discovery of the CMB was a watershed moment in the history of cosmology, providing direct evidence of the universe's hot and dense past.

2.1 Theoretical Predictions of the CMB

In the 1940s, George Gamow and his collaborators, Ralph Alpher and Robert Herman, developed the hot Big Bang model, which predicted that the early universe was filled with radiation. They calculated that this radiation should still be present in the universe, albeit much cooler and redshifted due to cosmic expansion. However, at the time, their work did not gain widespread attention, and the existence of the CMB remained unconfirmed.

2.2 The Accidental Discovery of the CMB

The CMB was first detected in 1964 by American astronomers Arno Penzias and Robert Wilson at Bell Laboratories. Their discovery was initially unintentional, as they were investigating a source of excess noise in their radio antenna, which turned out to be the CMB radiation. The significance of their discovery was soon recognized, and Penzias and Wilson were awarded the Nobel Prize in Physics in 1978 for their work.

3.Observations and Properties of the Cosmic Microwave Background

The CMB has been extensively studied and mapped since its discovery, providing a wealth of information about the early universe and the large-scale structure of the cosmos.

3.1 The Blackbody Spectrum of the CMB

The CMB has a nearly perfect blackbody spectrum, which means that its intensity as a function of frequency closely matches the theoretical curve for a blackbody at a temperature of 2.725 Kelvin. This blackbody spectrum is a key feature of the CMB, as it indicates that the radiation originated from a thermal equilibrium state in the early universe.

3.2 Anisotropies and the Seeds of Cosmic Structure

While the CMB is remarkably uniform across the sky, it does exhibit tiny temperature fluctuations, or anisotropies, on the order of one part in 100,000. These anisotropies are the imprints of small density fluctuations in the early universe, which eventually grew into the large-scale structures such as galaxies and galaxy clusters that we see today.

The study of CMB anisotropies has provided crucial insights into the universe's geometry, composition, and history. They have also helped to constrain cosmological models and parameters, such as the age of the universe and the densities of dark matter and dark energy.

3.3 CMB Polarization

In addition to its temperature anisotropies, the CMB also exhibits a polarization pattern, which is caused by the scattering of photons by free electrons during the epoch of recombination. CMB polarization provides an independent probe of the universe's history and structure and can help to test and refine our understanding of cosmic inflation, the formation of large-scale structure, and the properties of neutrinos.

4.Cosmic Microwave Background Observations and Experiments

Since the discovery of the CMB, numerous experiments have been conducted to map and analyze its properties, leading to significant advances in our understanding of the universe.

4.1 The Cosmic Background Explorer (COBE)

Launched in 1989, the Cosmic Background Explorer (COBE) satellite was the first major mission dedicated to studying the CMB. COBE made two groundbreaking discoveries: it confirmed the blackbody spectrum of the CMB and detected the temperature anisotropies for the first time. These findings provided strong support for the hot Big Bang model and the existence of primordial density fluctuations.

4.2 The Wilkinson Microwave Anisotropy Probe (WMAP)

The Wilkinson Microwave Anisotropy Probe (WMAP) was launched in 2001 to build upon COBE's findings by providing higher resolution maps of the CMB anisotropies. WMAP's data has been instrumental in refining cosmological parameters and models, helping to establish the current standard model of cosmology, known as the ΛCDM model. WMAP's observations have also provided strong evidence for cosmic inflation and have confirmed that the universe is geometrically flat.

4.3 The Planck Satellite

The Planck satellite, launched in 2009, represents the most recent and comprehensive CMB mission to date. Planck has produced the most detailed and precise maps of the CMB anisotropies and polarization, further constraining cosmological parameters and models. Planck's data has also allowed for the study of various secondary effects in the CMB, such as the Sunyaev-Zel'dovich effect and gravitational lensing, which provide additional probes of cosmic structure and history.

The Cosmic Microwave Background radiation is a cornerstone of modern cosmology, providing a unique window into the early universe and the large-scale structure of the cosmos. Its discovery and subsequent study have solidified the Big Bang Theory as the prevailing model of the universe's origin and evolution, and its properties have helped to shape our understanding of the universe's geometry, composition, and history.

As future experiments and missions continue to probe the CMB with even greater precision and sensitivity, we can expect to gain further insights into the mysteries of the universe, such as the nature of dark matter and dark energy, the properties of neutrinos, and the details of cosmic inflation. The Cosmic Microwave Background radiation will undoubtedly remain a vital source of information and discovery in the ongoing quest to unravel the secrets of the cosmos.

15

Dark Matter and Dark Energy

Dark matter and dark energy are two of the most enigmatic and intriguing components of the universe. They have never been directly observed, yet their effects on cosmic structures and the expansion of the universe provide compelling evidence for their existence. Together, they account for approximately 95% of the mass-energy content of the universe, with dark matter comprising about 27% and dark energy about 68%. In this chapter, we will explore the nature, evidence, and implications of these mysterious phenomena and the ongoing efforts to understand and uncover their properties.

1. Dark Matter: The Invisible Scaffold of the Cosmos

Dark matter is a form of matter that interacts very weakly with electro-magnetic radiation, rendering it nearly invisible to telescopes and other instruments that detect light. Despite this elusive nature, dark matter plays a crucial role in the formation and evolution of cosmic structures.

1.1 Evidence for Dark Matter

The existence of dark matter is inferred from its gravitational effects on visible matter, such as stars and galaxies, as well as on the large-scale structure of the universe.

1.1.1 Galactic Rotation Curves

The first evidence for dark matter came from the study of the rotation curves of galaxies, which plot the orbital velocities of stars and gas as a function of their distance from the galactic center. According to Newtonian gravity, the velocities should decrease with distance, but observations have shown that they remain roughly constant, implying the presence of a large amount of unseen mass in the form of dark matter.

1.1.2 Gravitational Lensing

Dark matter also reveals itself through gravitational lensing, a phenomenon where the gravity of massive objects bends and distorts the path of light. Observations of lensing effects in galaxy clusters, such as the Bullet Cluster, have shown a clear separation between the visible matter (stars and gas) and the inferred mass distribution, providing strong evidence for the existence of dark matter.

1.1.3 The Cosmic Microwave Background and Large-Scale Structure

Dark matter's effects on the anisotropies of the Cosmic Microwave Background (CMB) radiation and the large-scale structure of the universe provide further evidence for its existence. The distribution of matter in the universe, as inferred from the CMB and galaxy surveys, is consistent with a dominant component of non-baryonic, cold dark matter.

1.2 The Search for Dark Matter Particles

The nature and composition of dark matter remain unknown, but various particle candidates have been proposed, including Weakly Interacting Massive Particles (WIMPs), axions, and sterile neutrinos.

1.2.1 Direct Detection Experiments

Direct detection experiments aim to observe the interactions of dark matter particles with ordinary matter in highly sensitive detectors located deep underground to shield them from cosmic rays and other background sources. Examples of such experiments include the Cryogenic Dark Matter Search (CDMS), the Large Underground Xenon (LUX) experiment, and the XENON1T experiment. To date, these experiments have not provided definitive evidence for dark matter particles, but they continue to improve their sensitivity and refine their search strategies.

1.2.2 Indirect Detection and Collider Experiments

Indirect detection experiments seek to observe the products of dark matter particle annihilations or decays in cosmic sources, such as the center of the Milky Way or nearby dwarf galaxies. Additionally, particle colliders like the Large Hadron Collider (LHC) attempt to produce dark matter particles in high-energy collisions. While these efforts have yet to yield a definitive discovery, they provide valuable constraints on dark matter properties and help to guide theoretical models.

2.Dark Energy: The Mysterious Force Driving Cosmic Acceleration

Dark energy is a mysterious form of energy that permeates all of space and drives the accelerating expansion of the universe. Its existence was first inferred from observations of Type Ia supernovae in distant galaxies, which revealed that the universe's expansion is accelerating rather than slowing down, as previously thought.

2.1 Evidence for Dark Energy

The evidence for dark energy comes from several independent observations, including:

2.1.1 Type Ia Supernovae

Type Ia supernovae are a standard candle, meaning that they have a known intrinsic brightness. By observing their apparent brightness, astronomers can determine their distance and measure the expansion rate of the universe at different epochs. In the late 1990s, two teams of astronomers independently discovered that distant Type Ia supernovae were fainter than expected, indicating that the universe's expansion rate is accelerating rather than decelerating.

2.1.2 Cosmic Microwave Background and Large-Scale Structure

The cosmic microwave background radiation and the large-scale distribution of matter in the universe also provide indirect evidence for dark energy. These observations are consistent with a universe dominated by dark energy, which would drive the accelerated expansion seen in the supernova studies.

2.2 The Nature of Dark Energy

The nature and origin of dark energy remain unknown, but several theories have been proposed to explain its existence.

2.2.1 The Cosmological Constant

The simplest explanation for dark energy is the cosmological constant, which was introduced by Albert Einstein in his theory of general relativity. The cosmological constant represents a vacuum energy density that would drive the accelerated expansion of the universe. While the cosmological constant is consistent with current observational data, it requires an extremely fine-tuned value to match observations, leading some to consider it unsatisfactory.

2.2.2 Modified Gravity

Another possible explanation for dark energy is modified gravity, where the laws of gravity are altered on cosmic scales. While modified gravity theories can explain the accelerating expansion without the need for dark energy, they face several challenges in matching observational data and explaining other cosmic phenomena.

2.2.3 Dynamical Dark Energy

Dynamical dark energy models propose that dark energy is not a constant but instead varies over time or space. Examples of such models include quintessence, phantom energy, and Chaplygin gas. These models offer the possibility of explaining the observed acceleration while also accounting for the coincidence problem, where the onset of cosmic acceleration appears to coincide with the present epoch.

3.Implications and Future Directions

The existence of dark matter and dark energy has significant implications for our understanding of the universe's history and structure.

3.1 Dark Matter and Galactic Formation

Dark matter plays a critical role in the formation and evolution of galaxies and galaxy clusters. The gravitational effects of dark matter help to bind visible matter and provide the scaffolding for the formation of cosmic structures. Understanding the properties and nature of dark matter is essential for modeling and predicting the structure and evolution of galaxies and the large-scale structure of the universe.

3.2 Dark Energy and the Fate of the Universe

The discovery of dark energy has significant implications for the fate of the universe. If the acceleration continues, it will eventually lead to a "Big Freeze," where the universe expands at an ever-increasing rate, leading to the dissipation of all matter and the eventual heat death of the universe. Alternatively, if the acceleration decreases or stops, the universe may collapse in a "Big Crunch" or enter a phase of eternal expansion or contraction.

3.3 Future Directions

The search for dark matter particles and the nature of dark energy remain active areas of research in cosmology and particle physics. Upcoming experiments, such as the Large Synoptic Survey Telescope (LSST), the Dark Energy Survey (DES), and the European Space Agency's Euclid mission, aim to provide further insights into the properties and nature of dark matter and dark energy. These experiments will provide more precise measurements of the cosmic microwave background radiation and the large-scale distribution of matter in the universe, which will help to constrain the properties of dark matter and dark energy.

Furthermore, new experiments such as the Dark Energy Spectroscopic Instrument (DESI), the Cherenkov Telescope Array (CTA), and the XENONnT experiment will probe the nature of dark matter particles with even greater sensitivity and explore new regions of parameter space.

In addition to experimental efforts, theoretical models continue to explore the nature and properties of dark matter and dark energy. The development of new theories and the refinement of existing ones will play a crucial role in uncovering the mysteries of these invisible components of the universe.

Dark matter and dark energy represent two of the most significant puzzles in modern physics and cosmology. While they have never been directly observed, their effects on cosmic structures and the expansion of the universe provide compelling evidence for their existence. The search for dark matter particles and the nature of dark energy are active areas of research, with ongoing efforts to probe their properties and uncover their secrets.

The discovery of dark matter and dark energy has fundamentally changed our understanding of the universe's history and structure, and their properties will continue to shape our models and predictions of the cosmos. As our technology and understanding of the universe continue to advance, we can expect to gain further insights into these enigmatic components of the universe and the mysteries they hold.

IV

The Uncertainty Principle

16

Introduction to the Uncertainty Principle

The Uncertainty Principle, also known as Heisenberg's Uncertainty Principle, is one of the fundamental principles of quantum mechanics. It states that it is impossible to measure both the position and momentum of a particle with arbitrary precision. In other words, the more precisely you know the position of a particle, the less precisely you can know its momentum, and vice versa. This principle has important implications for the behavior of matter on the atomic and subatomic level, and has helped shape our understanding of the quantum world.

The Uncertainty Principle was first proposed by Werner Heisenberg in 1927, as a way of explaining certain puzzling experimental results. Heisenberg had been studying the behavior of electrons in atoms, and had found that it was impossible to measure their position and momentum simultaneously with perfect accuracy. This was a surprising result, as it contradicted the classical laws of physics, which suggested that both position and momentum could be measured with arbitrary precision.

Heisenberg's uncertainty principle is usually expressed mathematically as follows:

$\Delta x * \Delta p \geq h/4\pi$

where Δx is the uncertainty in position, Δp is the uncertainty in momentum, and h is Planck's constant, which is a fundamental constant of nature. The principle states that the product of these uncertainties must be greater than or equal to $h/4\pi$. In other words, the more accurately you measure the position of a particle, the larger the uncertainty in its momentum, and vice versa.

The Uncertainty Principle has important implications for the behavior of matter on the atomic and subatomic level. For example, it implies that electrons in atoms do not orbit the nucleus in well-defined paths, as was previously thought. Instead, the electrons occupy what are called "orbitals," which are regions of space where there is a high probability of finding the electron. The Uncertainty Principle implies that the exact position and momentum of the electron cannot be known simultaneously, so it is impossible to say exactly where the electron is at any given moment.

The Uncertainty Principle also has important implications for the behavior of particles at very small scales. For example, it implies that particles can spontaneously appear and disappear from empty space, as long as they do so quickly enough that their energy and duration satisfy the uncertainty principle. This is known as quantum fluctuation, and it has been observed experimentally in a variety of contexts.

One of the most famous applications of the Uncertainty Principle is in the measurement of the position and momentum of particles. According to the principle, it is impossible to measure both the position and momentum of a particle with arbitrary precision. This means that there is a fundamental limit to how accurately we can measure the properties of particles on the atomic and subatomic level. In practice, this limit is typically much larger than the experimental uncertainties that we encounter in the laboratory, so the Uncertainty Principle does not prevent us from making accurate measurements of many physical properties.

However, there are certain situations where the Uncertainty Principle plays a more important role. For example, in experiments involving very cold atoms, the uncertainty in their momentum can be very small, which means that their position becomes more uncertain. This can make it difficult to measure their position accurately, and can lead to effects like quantum tunneling, where particles can pass through barriers that they would not be able to pass through according to classical physics.

Another important implication of the Uncertainty Principle is in the interpretation of quantum mechanics. In classical physics, particles are thought of as having definite properties like position and momentum, and these properties can be measured with arbitrary precision. However, in quantum mechanics, particles are described by wave functions, which represent probabilities of finding the particle in a particular state. The Uncertainty Principle implies that the wave function cannot be known with arbitrary precision, so it is impossible to say exactly where the particle is at any given moment. This has led to various interpretations of quantum mechanics, including the Copenhagen interpretation, which holds that particles do not have definite properties until they are observed, and the many-worlds interpretation, which holds that the wave function represents the branching of reality into multiple parallel universes.

The Uncertainty Principle also has important implications for the development of technology, particularly in the field of quantum computing. Quantum computers rely on the principles of quantum mechanics to perform calculations that would be impossible for classical computers. However, the Uncertainty Principle imposes fundamental limits on the accuracy of measurements and computations at the quantum level. This means that quantum computers must be designed in such a way as to minimize the effects of uncertainty and noise, which can limit their performance.

The Uncertainty Principle is one of the fundamental principles of quantum mechanics, which states that it is impossible to measure both the position and momentum of a particle with arbitrary precision. This principle has important implications for the behavior of matter on the atomic and subatomic level, and has helped shape our understanding of the quantum world. The Uncertainty Principle has also played an important role in the interpretation of quantum mechanics, and has implications for the development of technology, particularly in the field of quantum computing. While the principle poses fundamental limits on our ability to measure and manipulate particles on the quantum level, it also opens up new possibilities for understanding and harnessing the power of the quantum world.

Despite the limitations posed by the Uncertainty Principle, scientists have developed techniques to minimize the effects of uncertainty and make more precise measurements at the quantum level. One such technique is called "squeezing," which involves manipulating the properties of a particle to reduce the uncertainty in one variable (e.g. position) at the expense of increasing the uncertainty in another variable (e.g. momentum). Another technique is called "entanglement," which involves creating pairs of particles that are so strongly correlated that the properties of one particle can be used to infer the properties of the other, even at a distance.

In recent years, the Uncertainty Principle has been the subject of much research and debate, particularly in relation to the measurement problem in quantum mechanics. The measurement problem arises from the fact that the act of measuring a particle can change its state, which makes it difficult to reconcile with the idea of particles having definite properties. Some physicists have suggested that the Uncertainty Principle may be a manifestation of deeper principles, such as the complementarity principle, which holds that particles can exhibit both wave-like and particle-like behavior depending on the context.

The Uncertainty Principle has also been linked to the concept of information in quantum mechanics. In classical physics, information can be measured with arbitrary precision, but in quantum mechanics, the Uncertainty Principle imposes fundamental limits on the accuracy of measurements and computations. This has led some physicists to speculate that information may be a fundamental aspect of the universe, and that the Uncertainty Principle may be related to the fundamental limits of information processing.

In summary, the Uncertainty Principle is a fundamental principle of quantum mechanics, which states that it is impossible to measure both the position and momentum of a particle with arbitrary precision. This principle has important implications for the behavior of matter on the atomic and subatomic level, and has helped shape our understanding of the quantum world. While the Uncertainty Principle poses fundamental limits on our ability to measure and manipulate particles on the quantum level, it also opens up new possibilities for understanding and harnessing the power of the quantum world. As scientists continue to explore the mysteries of the quantum world, the Uncertainty Principle will likely play a central role in shaping our understanding of the universe.

17

Quantum Mechanics and Wave-Particle Duality

Quantum mechanics is the branch of physics that describes the behavior of matter and energy on the atomic and subatomic level. It provides a framework for understanding phenomena that cannot be explained by classical physics, and it has led to many important discoveries and technological advances. One of the key concepts in quantum mechanics is wave-particle duality, which describes the dual nature of particles and waves in the quantum world. In this essay, we will explore the concept of wave-particle duality and its implications for our understanding of the quantum world.

Wave-particle duality is the idea that particles can exhibit both wave-like and particle-like properties, depending on the context in which they are observed. This means that particles can be described both as localized objects with definite positions and momenta, and as wave-like entities with properties like frequency and wavelength. This concept was first proposed by Louis de Broglie in 1924, and it was later developed by other physicists like Werner Heisenberg and Erwin Schrödinger.

One of the key experiments that demonstrated the wave-like nature of particles is the double-slit experiment. In this experiment, a beam of particles (usually electrons or photons) is directed at a screen with two slits. Behind the screen, a detector measures the intensity of the particles at different points on a screen. If particles were strictly localized objects, we would expect to see two distinct patterns of intensity behind the two slits. However, what is observed is an interference pattern, which is characteristic of waves.

The interference pattern is explained by the wave nature of the particles. When the beam of particles passes through the two slits, it diffracts and produces waves that interfere with each other. The interference produces regions of high and low intensity on the screen, creating the interference pattern. This experiment demonstrates that particles can behave like waves, and that they can interfere with themselves.

The double-slit experiment is just one example of the many ways in which the wave-particle duality manifests in the quantum world. Other experiments, like the Stern-Gerlach experiment, which measures the spin of particles, also demonstrate the dual nature of particles. In this experiment, a beam of particles is passed through a magnetic field, which causes the particles to align themselves along a particular axis. The resulting pattern of particles on a screen is consistent with the wave-like properties of the particles.

The wave-particle duality has important implications for our understanding of the quantum world. One of the most significant implications is that particles can exist in multiple states simultaneously. This idea is captured by the concept of superposition, which states that a particle can be in multiple states at once, with a probability of being observed in each state. For example, an electron can be in a superposition of states where it has spin up and spin down, and it is only when it is measured that it collapses into one of the two states.

The concept of superposition is fundamental to quantum mechanics, and it has led to many important technological advances. For example, the development of quantum computers relies on the ability to manipulate particles in a superposition of states. The ability to manipulate particles in a superposition of states also has important implications for quantum cryptography, which is a method of encrypting data that relies on the principles of quantum mechanics.

The wave-particle duality also has important implications for our understanding of the fundamental nature of matter and energy. In classical physics, particles are described as localized objects with definite properties like position and momentum. However, in the quantum world, particles are described as wave-like entities with properties like frequency and wavelength. This suggests that the fundamental nature of matter and energy may be different than what we observe in classical physics.

Another implication of the wave-particle duality is that it challenges our ability to observe and measure particles in the quantum world. In classical physics, it is possible to measure the position and momentum of a particle with arbitrary precision, but in the quantum world, the uncertainty principle states that it is impossible to measure both properties simultaneously with arbitrary precision. This means that the act of measurement can alter the state of the particle, making it difficult to know its true properties.

The wave-particle duality also plays a central role in the interpretation of quantum mechanics. There are many different interpretations of quantum mechanics, each with its own way of explaining the behavior of particles in the quantum world. One of the most popular interpretations is the Copenhagen interpretation, which holds that particles do not have definite properties until they are observed. According to this interpretation, the wave function of a particle represents the probability of finding the particle in a particular state when it is observed.

Another interpretation of quantum mechanics is the many-worlds interpretation, which holds that particles exist in multiple states simultaneously, and that the act of measurement causes the universe to split into multiple parallel universes, each with a different outcome. This interpretation has gained popularity in recent years, and it has led to much discussion and debate among physicists and philosophers.

The wave-particle duality is a fundamental concept in quantum mechanics, which describes the dual nature of particles and waves in the quantum world. This concept has important implications for our understanding of the behavior of matter and energy on the atomic and subatomic level, and it has led to many important discoveries and technological advances. The wave-particle duality challenges our understanding of the fundamental nature of matter and energy, and it has led to many different interpretations of quantum mechanics. While the wave-particle duality may seem strange and counterintuitive, it is an essential concept for understanding the mysteries of the quantum world.

The wave-particle duality has been the subject of much research and debate in the field of quantum mechanics. Some physicists have proposed that the wave-particle duality may be related to deeper principles, such as the complementarity principle, which holds that particles can exhibit both wave-like and particle-like behavior depending on the context. Others have suggested that the wave-particle duality may be a manifestation of the fundamental limitations of our ability to observe and measure particles in the quantum world.

Despite the challenges posed by the wave-particle duality, scientists have developed techniques for observing and manipulating particles in the quantum world. For example, the development of techniques like quantum entanglement and quantum superposition have enabled scientists to manipulate particles in ways that were previously thought impossible. These techniques have led to many important discoveries and technological advances, including the development of quantum computers and quantum cryptography.

The wave-particle duality has also played an important role in the development of modern physics. The discovery of the wave-particle duality helped to pave the way for the development of quantum mechanics, which has revolutionized our understanding of the behavior of matter and energy on the atomic and subatomic level. The principles of quantum mechanics have also led to the development of many new technologies, including lasers, transistors, and superconductors.

The wave-particle duality is a fundamental concept in quantum mechanics, which describes the dual nature of particles and waves in the quantum world. This concept has important implications for our understanding of the behavior of matter and energy on the atomic and subatomic level, and it has led to many important discoveries and technological advances. While the wave-particle duality may seem strange and counterintuitive, it is an essential concept for understanding the mysteries of the quantum world, and it will likely continue to play a central role in the development of modern physics in the years to come.

18

Werner Heisenberg and the Uncertainty Principle

Werner Heisenberg was a German physicist who made important contributions to the development of quantum mechanics, including the formulation of the Uncertainty Principle. Heisenberg's work revolutionized our understanding of the behavior of matter and energy on the atomic and subatomic level, and it has had important implications for the development of modern physics and technology. In this essay, we will explore the life and work of Werner Heisenberg, and the significance of the Uncertainty Principle in the development of quantum mechanics.

Early Life and Education

Werner Heisenberg was born on December 5, 1901, in Würzburg, Germany. He was the son of a professor of medieval and modern philology, and he grew up in a family that placed a high value on education and intellectual pursuits. Heisenberg showed an early interest in science, and he was particularly fascinated by the work of Albert Einstein and Max Planck.

Heisenberg studied physics at the University of Munich, where he received his doctorate in 1923. His dissertation was on the theory of turbulence, and it demonstrated his early interest in the mathematical foundations of physics. After completing his doctorate, Heisenberg spent a year studying physics in Copenhagen, where he worked with Niels Bohr, one of the leading figures in the development of quantum mechanics.

Development of Quantum Mechanics

Heisenberg's work in Copenhagen was instrumental in the development of quantum mechanics. At the time, physicists were struggling to understand the behavior of electrons in atoms, and there was no theoretical framework that could explain their behavior. Heisenberg worked closely with Bohr and other physicists to develop a new theory that could explain the behavior of electrons in atoms.

Heisenberg's breakthrough came in 1925, when he formulated a new approach to quantum mechanics that would eventually become known as matrix mechanics. Matrix mechanics was a mathematical framework that allowed physicists to calculate the behavior of particles in the quantum world, and it was based on the concept of wave-particle duality, which we discussed in a previous essay.

Matrix mechanics was a revolutionary development in the history of physics, and it paved the way for the development of modern quantum mechanics. Heisenberg's work was soon followed by the development of other approaches to quantum mechanics, including Schrödinger's wave mechanics and Dirac's relativistic quantum mechanics.

The Uncertainty Principle

One of Heisenberg's most important contributions to the development of quantum mechanics was the formulation of the Uncertainty Principle. The Uncertainty Principle states that it is impossible to measure both the position and momentum of a particle with arbitrary precision. This means that the more precisely we know the position of a particle, the less precisely we know its momentum, and vice versa.

The Uncertainty Principle was first proposed by Heisenberg in 1927, and it emerged from his attempts to explain the behavior of electrons in atoms. Prior to the development of quantum mechanics, it was thought that electrons moved in well-defined orbits around the nucleus of an atom. However, Heisenberg's observations suggested that the position and momentum of an electron could not both be measured with arbitrary precision. This led Heisenberg to propose the Uncertainty Principle as a fundamental limit on our ability to measure and understand the behavior of particles in the quantum world.

The Uncertainty Principle has important implications for our understanding of the behavior of matter on the atomic and subatomic level. It suggests that particles do not have definite properties until they are observed, and that the act of measurement can alter the state of a particle. This idea is central to the interpretation of quantum mechanics, and it has led to much discussion and debate among physicists and philosophers.

The Uncertainty Principle also has important implications for the development of technology, particularly in the field of quantum computing. Quantum computers rely on

manipulating particles in a superposition of states, which is made possible by the Uncertainty Principle. The ability to manipulate particles in a superposition of states has important implications for cryptography, where it can be used to create unbreakable codes and secure communication systems.

The Uncertainty Principle has also been linked to the concept of information in quantum mechanics. In classical physics, information can be measured with arbitrary precision, but in quantum mechanics, the Uncertainty Principle imposes fundamental limits on the accuracy of measurements and computations. This has led some physicists to speculate that information may be a fundamental aspect of the universe, and that the Uncertainty Principle may be related to the fundamental limits of information processing.

Later Life and Contributions

Heisenberg continued to make important contributions to the field of physics throughout his career. In 1932, he discovered the principle of noncommutativity, which states that the order in which operators are applied can affect the outcome of calculations in quantum mechanics. This principle is important for the development of many quantum mechanical techniques and is a fundamental principle of the theory.

Heisenberg's work in physics also extended to the study of nuclear physics. He was one of the key figures in the development of the first nuclear reactor, which was constructed by the German physicist Otto Hahn in 1938. Heisenberg also played a role in the development of the atomic bomb during

World War II, although he ultimately failed to develop a working bomb for the German government.

After the war, Heisenberg continued his work in physics, and he played an important role in the development of the field of particle physics. He also made significant contributions to the philosophy of science, where he argued for a holistic approach to understanding the universe. He was awarded the Nobel Prize in Physics in 1932 for his contributions to the development of quantum mechanics.

In the latter part of his career, Heisenberg became increasingly interested in the philosophical implications of quantum mechanics. He argued that the uncertainty principle represented a fundamental limit on our ability to understand the universe, and that there were limits to the knowledge that we could obtain through observation and experimentation. He also argued for a more holistic approach to understanding the universe, which emphasized the interconnectedness of all things.

The Uncertainty Principle is a fundamental concept in quantum mechanics, which states that it is impossible to measure both the position and momentum of a particle with arbitrary precision. The principle is often expressed mathematically using the following formula:

$\Delta x * \Delta p >= h/4\pi$

where Δx is the uncertainty in the position of the particle, Δp is the uncertainty in the momentum of the particle, h is Planck's constant, and π is the mathematical constant pi.

This formula states that the product of the uncertainties in the position and momentum of a particle must be greater than or equal to $h/4\pi$. In other words, the more precisely we know the position of a particle, the less precisely we can know its momentum, and vice versa. This means that there is a fundamental limit to our ability to measure and understand the behavior of particles in the quantum world.

Werner Heisenberg developed the Uncertainty Principle in 1927 as a fundamental limit on our ability to measure and manipulate particles on the quantum level. The principle has important implications for our understanding of the behavior of matter and energy on the atomic and subatomic level, and it has led to many important discoveries and technological advances. While the principle poses fundamental limits on our ability to measure and manipulate particles on the quantum level, it also opens up new possibilities for understanding and harnessing the power of the quantum world.

Werner Heisenberg was one of the most important physicists of the 20th century, and his contributions to the development of quantum mechanics have had a profound impact on our understanding of the behavior of matter and energy on the atomic and subatomic level. Heisenberg's work revolutionized our understanding of the quantum world, and it has led to many important discoveries and technological advances. The Uncertainty Principle, in particular, is a fundamental concept in quantum mechanics, which states that it is impossible to measure both the position and momentum of a particle with arbitrary precision. While the Uncertainty Principle poses fundamental limits on our ability to measure and manipulate particles on the quantum level, it also opens up new possibilities for understanding and harnessing the power of the quantum world.

19

Quantum Entanglement and Nonlocality

Quantum entanglement is one of the most intriguing and mysterious phenomena in the field of physics. It refers to a quantum mechanical state in which two or more particles become correlated in such a way that the state of one particle cannot be described independently of the state of the other particle, even if the particles are separated by large distances. This phenomenon has important implications for our understanding of the nature of reality, and it has led to many important discoveries and technological advances, including the development of quantum computers and secure communication systems.

In this essay, we will explore the concept of quantum entanglement and its relationship to nonlocality, which is another fundamental concept in the field of quantum mechanics. We will also discuss some of the experimental evidence for quantum entanglement and its applications in the development of modern technology.

Quantum Entanglement

Quantum entanglement is a phenomenon that arises in the quantum mechanical description of particles. In quantum mechanics, particles are described by a wave function, which contains all the information about the possible states of the particle. When two or more particles become entangled, their wave functions become correlated in such a way that the state of one particle cannot be described independently of the state of the other particle.

One of the key features of quantum entanglement is that it can exist over large distances, even across the entire universe. This means that the state of an entangled particle can be affected instantaneously by the state of another entangled particle, regardless of the distance between them. This violates the principle of locality, which states that physical systems cannot be affected by anything that is not in their immediate vicinity.

The concept of quantum entanglement was first proposed by Albert Einstein, Boris Podolsky, and Nathan Rosen in a paper published in 1935. The paper was intended to show that quantum mechanics was an incomplete theory, and that there must be some hidden variables that could account for the apparent nonlocality of entangled particles. However, subsequent experiments have shown that the predictions of quantum mechanics are correct, and that entanglement is a fundamental feature of the quantum world.

Nonlocality

Nonlocality is another fundamental concept in the field of quantum mechanics, which refers to the ability of particles to become correlated in such a way that the state of one particle can affect the state of another particle, even if the particles are separated by large distances. This violates the principle of locality, which states that physical systems cannot be affected by anything that is not in their immediate vicinity.

Nonlocality is closely related to the concept of quantum entanglement, and the two concepts are often used interchangeably. However, it is important to note that not all quantum systems that exhibit nonlocality are entangled, and not all entangled systems exhibit nonlocality.

One of the key features of nonlocality is that it allows for the possibility of instantaneous communication over large distances. This has important implications for the development of secure communication systems, where it can be used to create unbreakable codes and prevent eavesdropping.

Experimental Evidence for Quantum Entanglement

The concept of quantum entanglement may seem strange and counterintuitive, but there is a growing body of experimental evidence that supports its existence. One of the most famous experiments is the Bell test, which was first proposed by physicist John Bell in 1964.

The Bell test is a series of experiments that test whether or not particles can be entangled in such a way that their properties are correlated over large distances. The experiment involves measuring the polarization of entangled photons that have been separated by large distances. The results of the experiment are compared to the predictions of classical physics, which would suggest that the properties of the photons are not correlated over large distances.

The results of the Bell test have consistently shown that the predictions of quantum mechanics are correct, and that entanglement is a real and fundamental feature of the quantum world. The experiments have also shown that the correlations between entangled particles are nonlocal, meaning that the state of one particle can be affected instantaneously by the state of another particle, even if the particles are separated by large distances.

Another experiment that provides evidence for quantum entanglement is the EPR (Einstein-Podolsky-Rosen) experiment, which was first proposed by Einstein, Podolsky, and Rosen in their 1935 paper on quantum entanglement. The experiment involves entangling two particles and then measuring their properties at a distance. The results of the experiment have been shown to violate the principle of locality and provide evidence for the existence of entanglement.

Applications of Quantum Entanglement

Quantum entanglement has important applications in the development of modern technology, particularly in the fields of computing and communication. One of the most promising applications of quantum entanglement is in the development of quantum computers, which use the principles of entanglement and superposition to perform calculations that are impossible on classical computers.

Another application of quantum entanglement is in the development of secure communication systems. Entangled particles can be used to create unbreakable codes and secure communication channels, which are immune to eavesdropping and other forms of interference.

Quantum entanglement has also been used to test the foundations of quantum mechanics and to explore the nature of reality. For example, the concept of entanglement has been used to explore the possibility of a hidden variable theory, which would explain the apparent nonlocality of entangled particles by positing the existence of hidden variables that determine the behavior of the particles. However, experimental evidence has shown that such theories are unlikely to be correct, and that entanglement is a real and fundamental feature of the quantum world.

Quantum entanglement is a fundamental phenomenon in the field of physics, which refers to the correlation between the states of two or more particles, even if they are separated by large distances. The concept of entanglement is closely related to the concept of nonlocality, which refers to the ability of particles to become correlated in such a way that the state of one particle can affect the state of another particle, even if the particles are separated by large distances.

There is a growing body of experimental evidence that supports the existence of quantum entanglement, and the phenomenon has important implications for our understanding of the nature of reality and the development of modern technology. Entanglement has been used to develop new technologies, including quantum computers and secure communication systems, and it has also been used to test the foundations of quantum mechanics and explore the nature of the universe.

While the concept of quantum entanglement may seem strange and counterintuitive, it has important implications for our understanding of the behavior of matter and energy on the atomic and subatomic level, and it will likely continue to play a central role in the development of modern physics and technology in the years to come.

20

The Copenhagen Interpretation and its Alternatives

The Copenhagen Interpretation is one of the most well-known interpretations of quantum mechanics, which was developed by a group of physicists including Niels Bohr and Werner Heisenberg in the 1920s and 1930s. The interpretation emphasizes the role of measurement in quantum mechanics, and it has been the subject of much debate and controversy in the field of physics. In this essay, we will explore the Copenhagen Interpretation and its alternatives, including the Many-Worlds Interpretation and the Bohmian Interpretation.

The Copenhagen Interpretation

The Copenhagen Interpretation is based on the idea that the wave function of a quantum system contains all the possible states of the system, and that the act of measurement collapses the wave function into a single state. In other words, the interpretation emphasizes the role of measurement in quantum mechanics, and it suggests that the act of measurement is responsible for creating the classical reality that we observe.

The interpretation is named after the city of Copenhagen, where Niels Bohr established the Institute of Theoretical Physics in the 1920s. Bohr and his colleagues, including Werner Heisenberg and Max Born, developed the interpretation in response to the strange and counterintuitive behavior of particles on the atomic and subatomic level, which seemed to defy the principles of classical physics.

One of the key features of the Copenhagen Interpretation is the idea of complementarity, which suggests that certain properties of particles, such as position and momentum, are complementary and cannot be measured simultaneously with arbitrary precision. This is related to the Uncertainty Principle, which states that there is a fundamental limit to our ability to measure both the position and momentum of a particle.

The interpretation also emphasizes the role of observer dependence in quantum mechanics, suggesting that the act of measurement is inherently subjective and that the results of measurements depend on the observer and the apparatus used to make the measurement. This has important implications for our understanding of the nature of reality, and it has led some physicists to argue that quantum mechanics is incomplete or that there may be hidden variables that determine the behavior of particles.

Many-Worlds Interpretation

The Many-Worlds Interpretation is an alternative interpretation of quantum mechanics, which was first proposed by physicist Hugh Everett in the 1950s. The interpretation suggests that every time a quantum measurement is made, the universe splits into multiple branches, each representing a different outcome of the measurement. In other words, the interpretation suggests that all possible outcomes of a quantum measurement occur in parallel universes.

The Many-Worlds Interpretation is based on the idea of unitary evolution, which suggests that the wave function of a quantum system evolves in a deterministic way, without the need for the wave function to collapse upon measurement. This interpretation implies that there is no observer dependence in quantum mechanics, and that the results of measurements are objective and independent of the observer.

One of the key features of the Many-Worlds Interpretation is the idea of decoherence, which suggests that interactions between a quantum system and its environment lead to the emergence of classical behavior. This implies that the wave function of a quantum system does not collapse upon measurement, but rather, it becomes entangled with the environment, leading to the emergence of classical behavior.

The Many-Worlds Interpretation has been the subject of much debate and controversy in the field of physics. While some physicists find the interpretation to be elegant and satisfying, others find it to be too speculative and lacking in empirical support.

Bohmian Interpretation

The Bohmian Interpretation, also known as the de Broglie-Bohm Interpretation, is another alternative interpretation of quantum mechanics, which was developed by physicist David Bohm in the 1950s. The interpretation suggests that particles have well-defined positions and velocities, and that the wave function of a quantum system serves

as a pilot wave that guides the motion of the particles.

The Bohmian Interpretation is based on the idea of non-locality, which suggests that the behavior of a quantum system is influenced by its environment, even if the environment is far away from the system. This implies that the wave function of a quantum system does not collapse upon measurement, but rather, it guides the motion of the particles in a way that is consistent with the results of measurements.

One of the key features of the Bohmian Interpretation is the idea of hidden variables, which suggests that there are additional variables that determine the behavior of particles in a way that is consistent with the predictions of quantum mechanics. This implies that quantum mechanics is a complete theory, and that there are no observer-dependent or probabilistic elements in the theory.

The Bohmian Interpretation has been the subject of much debate and controversy in the field of physics. While some physicists find the interpretation to be elegant and satisfying, others find it to be too speculative and lacking in empirical support.

Comparison of Interpretations

Each interpretation of quantum mechanics offers a different perspective on the behavior of particles on the atomic and subatomic level, and each interpretation has its own strengths and weaknesses. The Copenhagen

Interpretation emphasizes the role of measurement in quantum mechanics, while the Many-Worlds Interpretation suggests that all possible outcomes of a measurement occur in parallel universes. The Bohmian Interpretation suggests that particles have well-defined positions and velocities, and that the wave function of a quantum system guides the motion of the particles.

One of the key differences between the interpretations is the role of observer dependence in quantum mechanics. The Copenhagen Interpretation suggests that the act of measurement is inherently subjective and that the results of measurements depend on the observer and the apparatus used to make the measurement. The Many-Worlds Interpretation, on the other hand, suggests that there is no observer dependence in quantum mechanics, and that the results of measurements are objective and independent of the observer. The Bohmian Interpretation suggests that there are hidden variables that determine the behavior of particles, and that quantum mechanics is a complete theory with no observer-dependent or probabilistic elements.

Another difference between the interpretations is the role of non-locality in quantum mechanics. The Many-Worlds Interpretation suggests that the behavior of particles is influenced by its environment, even if the environment is far away from the system. The Bohmian Interpretation also suggests that particles are influenced by their environment, but it does so through the concept of hidden variables. The Copenhagen Interpretation, on the other hand, does not explicitly address the issue of non-locality.

The Copenhagen Interpretation is one of the most well-known interpretations of quantum mechanics, and it emphasizes the role of measurement in the theory. The Many-Worlds Interpretation and the Bohmian Interpretation are two alternative interpretations, each with its own strengths and weaknesses. While each interpretation offers a different perspective on the behavior of particles on the atomic and subatomic level, none of them can be conclusively proven or disproven by experimental evidence. As such, the debate over the correct interpretation of quantum mechanics is likely to continue for years to come.

V

Elementary Particles and the Forces of Nature

21

Introduction to Elementary Particles

Elementary particles are the building blocks of matter in the universe. They are the most basic and fundamental constituents of all matter and are categorized into two main groups, fermions and bosons. Fermions are particles that make up matter, such as electrons, quarks, and neutrinos, while bosons are particles that carry the fundamental forces of nature, such as photons, W and Z bosons, and gluons. In this essay, we will explore the world of elementary particles and their properties.

Fermions

Fermions are particles that make up matter and are divided into two types: quarks and leptons. Quarks are the building blocks of protons and neutrons, which are the fundamental particles that make up the nucleus of an atom. There are six types of quarks, known as up, down, charm, strange, top, and bottom. Leptons, on the other hand, are fundamental particles that do not experience the strong force that binds quarks together. There are three types of leptons, known as electrons, muons, and taus.

Each type of fermion has a corresponding antiparticle, which has the opposite charge and spin. For example, the antiparticle of an electron is a positron, which has the same mass as an electron but has a positive charge. When a particle and its corresponding antiparticle come into contact, they annihilate each other and release energy in the form of gamma rays.

Fermions obey the Pauli exclusion principle, which states that no two fermions can occupy the same quantum state simultaneously. This principle gives rise to the concept of spin, which is a fundamental property of fermions that determines their behavior in magnetic fields. Fermions with half-integer spins, such as electrons, are known as spin-1/2 particles.

Bosons

Bosons are particles that carry the fundamental forces of nature, such as electromagnetism, the weak force, and the strong force. There are four types of bosons: photons, W and Z bosons, and gluons.

Photons are particles that carry the electromagnetic force and are responsible for the interaction of charged particles, such as electrons and protons. Photons are massless and travel at the speed of light.

W and Z bosons are particles that carry the weak force, which is responsible for the decay of particles. W bosons come in two types, known as W+ and W-, which carry positive and negative charges, respectively. Z bosons, on the other hand, are neutral and do not carry a charge.

Gluons are particles that carry the strong force, which is responsible for binding quarks together to form protons and neutrons. Gluons come in eight types, known as red, blue, and green, and their corresponding antiparticles.

Unlike fermions, bosons do not obey the Pauli exclusion principle and can occupy the same quantum state simultaneously. This gives rise to the concept of superposition, which is a fundamental property of bosons that allows them to form coherent states, such as the laser.

Properties of Elementary Particles

Elementary particles have a number of properties that are used to describe their behavior and interactions. Some of the key properties of elementary particles include:

1. Mass: The mass of a particle is a measure of its inertia and is an important property that determines its behavior in interactions.
2. Charge: The charge of a particle determines how it interacts with other charged particles, such as electrons and protons.
3. Spin: The spin of a particle is a fundamental property that determines its behavior in magnetic fields.

4. Flavor: The flavor of a particle is a property that distinguishes between different types of particles, such as up and down quarks.Color: The color of a particle is a property that distinguishes between different types of quarks and is related to the strong force that binds them together.

5. Isospin: The isospin of a particle is a property that distinguishes between different types of particles that are affected by the weak force, such as protons and neutrons.

These properties are used to classify and describe the behavior of elementary particles and their interactions.

Interactions between Elementary Particles

Elementary particles interact with each other through the fundamental forces of nature, which are mediated by bosons. The four fundamental forces are:

1. Electromagnetism: The force responsible for the interaction of charged particles, such as electrons and protons. This force is mediated by photons.

2. Weak force: The force responsible for the decay of particles, such as beta decay. This force is mediated by W and Z bosons.

3. Strong force: The force responsible for binding quarks together to form protons and neutrons. This force is mediated by gluons.

4. Gravity: The force responsible for the attraction of masses. This force is mediated by hypothetical particles known as gravitons.

The interactions between particles can be described using quantum field theory, which is a mathematical framework that combines quantum mechanics and special relativity. In this framework, particles are described as excitations of quantum fields, and interactions between particles are described as exchanges of bosons.

Applications of Elementary Particles

Elementary particles have important applications in many areas of modern technology, including medicine, energy, and telecommunications. One of the most important applications of elementary particles is in the development of particle accelerators, which are used to study the behavior of particles on the atomic and subatomic level.

Particle accelerators have been used to discover many new particles, including the Higgs boson, which was discovered in 2012 at the Large Hadron Collider in Switzerland. The discovery of the Higgs boson confirmed the existence of the Higgs field, which is a fundamental field that gives particles mass.

Another important application of elementary particles is in the development of medical imaging technologies, such as PET (positron emission tomography) scanners, which use positrons to produce images of the body. Positrons are the antiparticles of electrons and are produced by the decay of radioactive isotopes.

Elementary particles also have important applications in energy, particularly in the development of nuclear power. Nuclear power plants use the energy released by the decay of radioactive isotopes to produce electricity. The behavior of particles on the atomic and subatomic level is also important for understanding the behavior of materials and the properties of matter.

Elementary particles are the building blocks of matter and the most basic and fundamental constituents of all matter. They are categorized into two main groups, fermions and bosons, and have a number of properties that are used to describe their behavior and interactions. The interactions between particles are mediated by the fundamental forces of nature, which are described using quantum field theory. Elementary particles have important applications in many areas of modern technology, including medicine, energy, and telecommunications, and the study of elementary particles is likely to continue to play a central role in the development of modern science and technology.

22

The Standard Model of Particle Physics

The Standard Model of particle physics is a theory that describes the behavior of elementary particles and their interactions through the fundamental forces of nature. It is one of the most successful and comprehensive theories in modern physics, and it has been confirmed by many experimental observations.

In this essay, we will explore the Standard Model of particle physics, its components, and its predictions.

Introduction

The Standard Model of particle physics is a mathematical framework that describes the behavior of elementary particles and their interactions. It is based on the principles of quantum mechanics and special relativity and is divided into two main components: the gauge theory of the strong, weak, and electromagnetic forces and the Higgs mechanism, which gives particles mass.

The theory was developed in the 1960s and 1970s by a group of physicists, including Sheldon Glashow, Abdus Salam, and Steven Weinberg, and it has been confirmed by many experimental observations.

Components of the Standard Model

The Standard Model is divided into two main components: the gauge theory of the strong, weak, and electromagnetic forces and the Higgs mechanism.

Gauge Theory of the Forces

The gauge theory of the forces describes the behavior of particles that interact through the strong, weak, and electromagnetic forces. The theory is based on the concept of gauge invariance, which states that the laws of physics should be independent of the choice of gauge.

The strong force is responsible for binding quarks together to form protons and neutrons, which are the building blocks of atomic nuclei. The strong force is mediated by the exchange of gluons, which are massless particles that carry a color charge. The strong force is described by the theory of quantum chromodynamics (QCD), which is a gauge theory that describes the behavior of quarks and gluons.

The weak force is responsible for the decay of particles, such as beta decay. The weak force is mediated by the exchange of W and Z bosons, which are massive particles that carry weak charges. The weak force is described by the theory of electroweak interaction, which is a gauge theory that describes the behavior of particles that interact through the weak and electromagnetic forces.

The electromagnetic force is responsible for the interaction of charged particles, such as electrons and protons. The electromagnetic force is mediated by the exchange of photons, which are massless particles that carry a charge. The electromagnetic force is described by the theory of quantum electrodynamics (QED), which is a gauge theory that describes the behavior of charged particles.

Higgs Mechanism

The Higgs mechanism is responsible for giving particles mass. The mechanism is based on the concept of spontaneous symmetry breaking, which occurs when a system is in a symmetric state but the ground state is not symmetric.

The Higgs mechanism involves the interaction of particles with a scalar field, known as the Higgs field. The interaction gives particles mass by slowing them down as they move through the field. The Higgs mechanism also predicts the existence of a massive particle, known as the Higgs boson, which was discovered in 2012 at the Large Hadron Collider in Switzerland.

Predictions of the Standard Model

The Standard Model makes a number of predictions about the behavior of particles and their interactions. Some of the key predictions include:

1. Existence of the Higgs Boson: The Standard Model predicts the existence of a massive particle, known as the Higgs boson, which was discovered in 2012 at the Large Hadron Collider in Switzerland.

Quark and Lepton Families: The Standard Model predicts the existence of three families of quarks and leptons, each with increasing mass. The first family includes up and down quarks, electrons, and neutrinos. The second family includes charm and strange quarks, muons, and muon neutrinos. The third family includes top and bottom quarks, taus, and tau neutrinos.

1. Neutrino Oscillations: The Standard Model predicts that neutrinos can change flavor as they travel through space, a phenomenon known as neutrino oscillation. This has been confirmed by many experimental observations.
2. W and Z Boson Masses: The Standard Model predicts the masses of the W and Z bosons, which have been confirmed by many experimental observations.
3. Charge Quantization: The Standard Model predicts that the electric charge of particles is quantized, meaning that it can only take on certain discrete values.
4. Unification of Forces: The Standard Model predicts that the strong, weak, and electromagnetic forces can be unified at high energies, a phenomenon known as grand unification. This has not yet been confirmed by experimental observations.

Challenges to the Standard Model

While the Standard Model has been confirmed by many experimental observations, it is not a complete theory of particle physics. There are a number of challenges to the Standard Model, including:

1. Dark Matter: The Standard Model does not explain the existence of dark matter, which is believed to make up approximately 27% of the universe.

2. Neutrino Masses: The Standard Model does not explain why neutrinos have mass, which has been confirmed by many experimental observations.

3. Hierarchy Problem: The Standard Model predicts the existence of a massive Higgs boson, but the mass of the Higgs boson is much lighter than what would be expected based on the interactions of particles in the theory.

4. CP Violation: The Standard Model does not explain why there is a difference in the behavior of matter and antimatter, known as CP violation.

5. Grand Unification: The Standard Model predicts that the strong, weak, and electromagnetic forces can be unified at high energies, but this has not yet been confirmed by experimental observations.

The Standard Model of particle physics is a comprehensive theory that describes the behavior of elementary particles and their interactions. The theory is based on the principles of quantum mechanics and special relativity and is divided into two main components: the gauge theory of the strong, weak, and electromagnetic forces and the Higgs mechanism, which gives particles mass. While the Standard Model has been confirmed by many experimental observations, there are a number of challenges to the theory, including the existence of dark matter, the origin of neutrino masses, the hierarchy problem, CP violation, and the unification of the fundamental forces. Future research in particle physics is likely to focus on resolving these challenges and developing a more complete theory of the behavior of matter and energy in the universe.

23

The Four Fundamental Forces

The four fundamental forces are the basic interactions that govern the behavior of matter and energy in the universe. These forces are the gravitational force, the electromagnetic force, the strong force, and the weak force. Each force has a different range and strength, and they interact in different ways with matter.

Gravitational Force

The gravitational force is the force of attraction between two objects with mass. It is the weakest of the four fundamental forces but has an infinite range. This force is responsible for holding galaxies and clusters of galaxies together.

The gravitational force is described by Isaac Newton's law of gravitation, which states that the force between two objects is proportional to the product of their masses and inversely proportional to the square of the distance between them. In other words, the force of gravity between two objects increases with their mass and decreases with the distance between them.

The gravitational force is also described by Albert Einstein's theory of general relativity, which describes gravity as the curvature of spacetime. According to this theory, objects with mass curve spacetime around them, and other objects moving in this curved spacetime follow a curved path, giving the appearance of a gravitational force.

Electromagnetic Force

The electromagnetic force is the force of attraction or repulsion between electrically charged particles. It is responsible for the behavior of charged particles, such as electrons and protons, and for the behavior of electromagnetic waves, such as light.

The electromagnetic force is described by James Clerk Maxwell's equations of electromagnetism, which describe how electric and magnetic fields interact with matter. These equations predict the existence of electromagnetic waves, which travel at the speed of light and include radio waves, microwaves, infrared radiation, visible light, ultraviolet radiation, X-rays, and gamma rays.

The electromagnetic force is responsible for holding atoms and molecules together and is responsible for the behavior of many everyday objects, such as magnets, electric motors, and generators.

Strong Force

The strong force is the force that binds protons and neutrons together in the nucleus of an atom. It is the strongest of the four fundamental forces but has a very short range, only acting over distances of approximately 10^{-15} meters.

The strong force is described by the theory of quantum chromodynamics (QCD), which describes the behavior of quarks and gluons. Quarks are the building blocks of protons and neutrons, and gluons are the particles that bind quarks together.

The strong force is responsible for the stability of atomic nuclei and is responsible for the energy produced by nuclear reactions, such as those that occur in the sun.

Weak Force

The weak force is the force responsible for the decay of particles, such as beta decay. It is the second weakest of the four fundamental forces and has a range of only approximately 10^{-18} meters.

The weak force is described by the theory of electroweak interaction, which describes the behavior of particles that interact through the weak and electromagnetic forces. The theory predicts the existence of W and Z bosons, which are the particles that mediate the weak force.

The weak force is responsible for the behavior of particles that decay through the weak force, such as neutrinos, and is responsible for the energy produced by nuclear reactions.

Interactions Between Fundamental Forces

The four fundamental forces interact with each other in different ways, and their interactions can be described using the principles of quantum field theory.

The electromagnetic force and the weak force are combined into a single electroweak force at high energies, such as those found in the early universe. This force is described by the theory of electroweak interaction.

The strong force and the electroweak force are believed to be unified at even higher energies, a phenomenon known as grand unification. However, this has not yet been confirmed by experimental observations.

Challenges to the Fundamental Forces

While the four fundamental forces have been extensively studied and confirmed by experimental observations, there are still challenges and unanswered questions about their behavior and interactions.

One major challenge is the existence of dark matter, which is believed to make up approximately 27% of the universe's total mass. Dark matter interacts only weakly with ordinary matter and has not been directly detected, leading to much speculation and research about its nature.

Another challenge is the problem of hierarchy, which relates to the large discrepancy between the strength of the gravitational force and the other three fundamental forces. This problem is related to the mass of the Higgs boson, which is responsible for giving particles mass. The Higgs boson is much lighter than what would be expected based on the interactions of particles in the theory.

Finally, the unification of the fundamental forces remains an ongoing area of research. While the electroweak force has been unified at high energies, the strong force has yet to be fully incorporated into a grand unified theory.

here are the mathematical formulas that describe the four fundamental forces:

1. Gravitational force: $F = G\,(m_1m_2/r^2)$ where F is the force of gravity, G is the gravitational constant, m_1 and m_2 are the masses of the two objects, and r is the distance between them.
2. Electromagnetic force: $F = k(q_1q_2/r^2)$ where F is the electromagnetic force, k is Coulomb's constant, q_1 and q_2 are the charges of the two objects, and r is the distance between them.
3. Strong force: the strong force is described by the theory of quantum chromodynamics (QCD), which uses mathematical equations and concepts such as quarks, gluons, and color charge.
4. Weak force: the weak force is described by the theory of electroweak interaction, which uses mathematical equations and concepts such as W and Z bosons and weak charge. One of the key equations in the theory is the Fermi interaction, which describes the decay of particles through the weak force:

$G\_F = h\_bar^2/(8m\_w^2)$ where G_F is the Fermi coupling constant, h_bar is the reduced Planck constant, and m_w is the mass of the W boson.

These equations are used by physicists to describe and predict the behavior of matter and energy in the universe, and they have been confirmed by many experimental observations.

The four fundamental forces are the basic interactions that govern the behavior of matter and energy in the universe. They are the gravitational force, the electromagnetic force, the strong force, and the weak force. Each force has a different range and strength and interacts with matter in different ways.

While the fundamental forces have been extensively studied and confirmed by experimental observations, there are still challenges and unanswered questions about their behavior and interactions. These include the existence of dark matter, the problem of hierarchy, and the unification of the fundamental forces. Future research in particle physics is likely to focus on resolving these challenges and developing a more complete theory of the behavior of matter and energy in the universe.

24

The Higgs Boson and the Higgs Field

The Higgs boson and the Higgs field are two of the most important concepts in modern particle physics. They are central to the Standard Model of particle physics and help explain how particles acquire mass. In this essay, we will explore the Higgs boson and the Higgs field, their discovery, and their significance in the field of particle physics.

Introduction

The Higgs boson and the Higgs field are two interrelated concepts that were first proposed in the 1960s by several physicists, including Peter Higgs, François Englert, and Robert Brout. Their work helped develop the Standard Model of particle physics, which describes the behavior of elementary particles and their interactions through the fundamental forces of nature.

The Higgs boson and the Higgs field are crucial to the Standard Model because they provide an explanation for the origin of mass. Prior to the development of the Higgs mechanism, it was not clear why some particles had mass and others did not. The Higgs mechanism helps explain why particles acquire mass and how this mass is generated.

The Higgs Boson

The Higgs boson is a massive particle that was first predicted to exist in the 1960s. It was named after Peter Higgs, one of the physicists who proposed its existence. The Higgs boson was discovered in 2012 at the Large

Hadron Collider (LHC) in Switzerland, following decades of theoretical and experimental work.

The discovery of the Higgs boson was a major breakthrough in particle physics. It confirmed the existence of the Higgs field, which is responsible for giving particles mass, and helped validate the Standard Model of particle physics.

The Higgs boson has a mass of approximately 125 gigaelectronvolts (GeV), making it one of the heaviest elementary particles known to exist. It has a very short lifetime and decays into other particles almost immediately after it is produced.

The Higgs Field

The Higgs field is a scalar field that permeates all of space. It is responsible for giving particles mass through the Higgs mechanism. The Higgs field was first proposed by Peter Higgs in 1964, and its existence was confirmed in 2012 following the discovery of the Higgs boson.

The Higgs field is a unique field in that it has a non-zero value even in empty space. This value, known as the vacuum expectation value, is responsible for the generation of mass in particles. When particles interact with the Higgs field, they slow down and become more massive.

The Higgs Mechanism

The Higgs mechanism is a theoretical framework that explains how particles acquire mass through interactions with the Higgs field. The mechanism involves the interaction of particles with the Higgs field, which slows them down as they move through the field. This slowing down results in an increase in the particle's mass.

The Higgs mechanism is based on the concept of spontaneous symmetry breaking, which occurs when a system is in a symmetric state but the ground state is not symmetric. In the case of the Higgs mechanism, the Higgs field is initially in a symmetric state, but as particles interact with the field, it spontaneously breaks the symmetry, resulting in the generation of mass.

The Higgs mechanism involves the interaction of particles with a scalar field, known as the Higgs field. The interaction gives particles mass by slowing them down as they move through the field. The Higgs mechanism also predicts the existence of a massive particle, known as the Higgs boson, which was discovered in 2012 at the LHC.

The Higgs mechanism has been extensively tested and confirmed by many experimental observations. It is a crucial component of the Standard Model of particle physics and has helped explain the origin of mass in particles.

Significance of the Higgs Boson and the Higgs Field

The discovery of the Higgs boson and the confirmation of the Higgs field have significant implications for our understanding of the universe. They provide an explanation for why particles have mass and how this mass is generated. This understanding helps us better understand the behavior of particles and the interactions between them.

The Higgs field also has important implications for our understanding of the early universe. It is believed that the Higgs field played a crucial role in the formation of the universe in the moments following the Big Bang. The Higgs field would have been present at very high temperatures and energies, and its interactions with other particles would have played a key role in the evolution of the early universe.

In addition, the Higgs field is closely related to other fundamental forces of nature, such as the weak force. The Higgs mechanism predicts the existence of a massive particle, known as the Higgs boson, which is responsible for mediating the weak force. The weak force is one of the four fundamental forces of nature, and its behavior is closely tied to the Higgs field.

Challenges to the Higgs Mechanism

While the Higgs mechanism has been extensively tested and confirmed by many experimental observations, there are still challenges and unanswered questions about its behavior and interactions.

One of the main challenges is the so-called hierarchy problem, which relates to the large discrepancy between the strength of the gravitational force and the other three fundamental forces. The Higgs boson is much lighter than what would be expected based on the interactions of particles in the theory, and this has led to much speculation and research about possible solutions to this problem.

Another challenge is the problem of dark matter, which is believed to make up approximately 27% of the universe's total mass. Dark matter interacts only weakly with ordinary matter and has not been directly detected, leading to much speculation and research about its nature.

Finally, the unification of the fundamental forces remains an ongoing area of research. While the electroweak force has been unified at high energies, the strong force has yet to be fully incorporated into a grand unified theory.

The Higgs boson and the Higgs field are two of the most important concepts in modern particle physics. They are central to the Standard Model of particle physics and help explain how particles acquire mass. The discovery of the Higgs boson in 2012 at the LHC was a major breakthrough in particle physics, and it confirmed the existence of the Higgs field.

The Higgs mechanism is a theoretical framework that explains how particles acquire mass through interactions with the Higgs field. It has been extensively tested and confirmed by many experimental observations and is a crucial component of the Standard Model.

While the Higgs mechanism has been extensively studied and confirmed by experimental observations, there are still challenges and unanswered questions about its behavior and interactions. These include the hierarchy problem, the problem of dark matter, and the unification of the fundamental forces. Future research in particle physics is likely to focus on resolving these challenges and developing a more complete theory of the behavior of matter and energy in the universe.

25

Beyond the Standard Model: Supersymmetry and String Theory

The Standard Model of particle physics has been extremely successful in describing the behavior of elementary particles and their interactions through the fundamental forces of nature. However, there are still unanswered questions and phenomena that are not fully explained by the Standard Model. In this essay, we will explore two theories that go beyond the Standard Model: supersymmetry and string theory.

Supersymmetry

Supersymmetry is a theoretical framework that extends the Standard Model by introducing a new symmetry between fermions and bosons. In the Standard Model, fermions and bosons are two distinct types of particles that behave differently under the fundamental forces. However, in supersymmetry, fermions and bosons are considered to be different manifestations of the same underlying particle, called a superpartner.

The idea of supersymmetry was first proposed in the 1970s by several physicists, including Julius Wess and Bruno Zumino. Supersymmetry provides a solution to several problems that are not fully explained by the Standard Model, including the hierarchy problem and the problem of dark matter.

The Hierarchy Problem

One of the main challenges to the Standard Model is the hierarchy problem, which relates to the large discrepancy between the strength of the gravitational force and the other three fundamental forces. The Higgs boson, which is responsible for giving particles mass, is much lighter than what would be expected based on the interactions of particles in the theory.

Supersymmetry provides a solution to the hierarchy problem by introducing new particles, called supersymmetric particles or sparticles, that cancel out the contributions of the Standard Model particles to the Higgs boson mass. The sparticles have the same quantum numbers as their Standard Model counterparts, but they differ in their spin, which allows them to cancel out the contributions of the Standard Model particles.

The Problem of Dark Matter

Another challenge to the Standard Model is the problem of dark matter, which is believed to make up approximately 27% of the universe's total mass. Dark matter interacts only weakly with ordinary matter and has not been directly detected, leading to much speculation and research about its nature.

Supersymmetry provides a solution to the problem of dark matter by introducing a stable supersymmetric particle, called the lightest supersymmetric particle (LSP), that could be a candidate for dark matter. The LSP is predicted to be weakly interacting, which would make it difficult to detect directly.

String Theory

String theory is a theoretical framework that attempts to unify all of the fundamental forces of nature into a single, coherent theory. In string theory, the fundamental building blocks of the universe are not point-like particles but rather one-dimensional strings that vibrate at different frequencies. These vibrations determine the properties of the particles that are observed in experiments.

The idea of string theory was first proposed in the late 1960s by several physicists, including Gabriele Veneziano and Leonard Susskind. String theory provides a possible solution to the problem of unifying the fundamental forces of nature, as it predicts that all particles and forces are manifestations of the vibrations of one-dimensional strings.

The Five String Theories

There are several different versions of string theory, collectively known as the five string theories. These include type I, type IIa, type IIb, heterotic SO(32), and heterotic E8xE8. Each of these theories is characterized by the properties of the strings, including their vibrational modes and the number of dimensions of space-time required for their consistency.

One of the key features of string theory is the prediction of extra dimensions of space-time. In addition to the three dimensions of space and one dimension of time that we experience in our everyday lives, string theory predicts the existence of six or seven additional dimensions that are curled up and compactified. These extra dimensions are believed to be responsible for the unification of the fundamental forces of nature, as they allow for the different forces to interact in a single, higher-dimensional space.

Challenges to String Theory

Despite its promise as a theory of everything, string theory faces several challenges and criticisms. One of the main criticisms of string theory is that it is currently impossible to test experimentally, as the energies required to observe the strings directly are far beyond the capabilities of current particle accelerators.

Another challenge is the problem of compactification, or the process of curling up the extra dimensions of space-time. There are many ways to compactify the extra dimensions, and each leads to a different set of predictions for the behavior of particles and forces. It is not clear which compactification scenario is correct, and this has led to much speculation and research about possible solutions to this problem.

In addition, string theory does not predict the exact properties of the particles and forces that we observe in experiments. Rather, it predicts a vast landscape of possible universes, each with its own set of physical laws and properties. It is not clear how to select the correct universe from this landscape, and this has led to much debate and speculation about the nature of the theory.

Supersymmetry and string theory are two theories that go beyond the Standard Model of particle physics. Supersymmetry introduces a new symmetry between fermions and bosons, and it provides a solution to several problems that are not fully explained by the Standard Model, including the hierarchy problem and the problem of dark matter.

String theory attempts to unify all of the fundamental forces of nature into a single, coherent theory. It predicts that the fundamental building blocks of the universe are one-dimensional strings that vibrate at different frequencies, and it predicts the existence of extra dimensions of space-time that allow for the unification of the forces.

While both theories are still under active research and development, they offer exciting possibilities for understanding the nature of the universe and the behavior of matter and energy at the most fundamental levels. Their potential to answer some of the most profound questions in physics, such as the nature of dark matter and the unification of the forces, makes them a subject of great interest to physicists and the public alike.

VI

Black Holes

26

Introduction to Black Holes

Black holes are some of the most fascinating and mysterious objects in the universe. They are regions in space where gravity is so strong that nothing, not even light, can escape. In this essay, we will explore the concept of black holes, their properties, and their role in the universe.

What is a Black Hole?

A black hole is a region of space where the gravitational pull is so strong that nothing can escape, not even light. This means that anything that enters a black hole is trapped forever, unable to escape its gravitational pull.

Black holes are formed when massive stars collapse under their own weight, creating a singularity, a point of infinite density and zero volume. The singularity is surrounded by an event horizon, which is the point of no return for anything that enters the black hole. Once something crosses the event horizon, it is impossible for it to escape the gravitational pull of the black hole.

Properties of Black Holes

Black holes are characterized by three main properties: mass, spin, and charge.

Mass: The mass of a black hole is the amount of matter that has collapsed into the singularity. The more massive a black hole is, the stronger its gravitational pull.

Spin: The spin of a black hole is a measure of how fast it is rotating. The faster a black hole is spinning, the more distorted space-time becomes near the black hole.

Charge: Black holes can also have an electric charge, which is a measure of the distribution of electric charge within the black hole. However, most black holes are believed to be neutral, meaning they have no electric charge.

Types of Black Holes

There are three main types of black holes: stellar, intermediate, and supermassive.

Stellar black holes are the most common type of black hole, and they are formed from the collapse of massive stars. They have masses ranging from a few times that of the sun to tens of times the mass of the sun.

Intermediate black holes are less common and have masses between 100 and 100,000 times the mass of the sun. It is not clear how intermediate black holes are formed, but they may be the result of the merging of smaller black holes.

Supermassive black holes are the most massive type of black hole, with masses ranging from millions to billions of times the mass of the sun. They are believed to be located at the centers of most galaxies, including our own Milky Way galaxy.

Observing Black Holes

Black holes themselves cannot be directly observed, as they do not emit any light or radiation. However, the effects of black holes can be observed indirectly through their interactions with other matter and their gravitational influence on nearby objects.

One way that black holes can be detected is through their effects on nearby stars. If a star is orbiting around a black hole, its orbit can be measured and used to infer the mass of the black hole.

Another way that black holes can be detected is through the emission of radiation from matter falling into the black hole. As matter falls towards the black hole, it heats up and emits X-rays and other forms of radiation. This radiation can be detected by telescopes and used to infer the presence of a black hole.

Importance of Black Holes

Black holes play a crucial role in the universe, both in terms of their effects on nearby objects and their role in the evolution of galaxies.

Black holes are believed to be responsible for the formation of galaxies, as their gravitational pull can cause gas and dust to collapse and form stars. Supermassive black holes are also believed to play a role in regulating the growth of galaxies by regulating the flow of gas into and out of the galaxy.

Black holes can also have a profound effect on nearby objects. If a star passes too close to a black hole, it can be torn apart by the strong gravitational forces, producing a bright burst of radiation known as a tidal disruption event. Black holes can also cause gravitational lensing, which is a distortion of light from distant objects as it passes near a massive object.

Black holes are also important in the study of general relativity and the nature of space-time. They provide a unique testing ground for the predictions of Einstein's theory of general relativity, as their strong gravitational fields can cause deviations from the predictions of classical physics.

Challenges and Unsolved Mysteries

While black holes have been extensively studied and observed, there are still many unanswered questions and mysteries surrounding them.

One of the main challenges is the nature of the singularity at the center of a black hole. According to our current understanding of physics, the singularity is a point of infinite density and zero volume, which violates the laws of physics as we understand them. It is not clear what happens to matter that falls into a black hole or how the singularity behaves.

Another challenge is the problem of information loss. According to classical physics, anything that falls into a black hole is lost forever, as it cannot escape the event horizon. However, this would violate the laws of quantum mechanics, which require that information cannot be lost. The resolution to this paradox is still an active area of research and debate.

Finally, the study of black holes and their properties has also led to the discovery of new and unexpected phenomena, such as black hole thermodynamics and the holographic principle. These areas of research are still in their early stages, but they offer exciting possibilities for further understanding the nature of black holes and their place in the universe.

Black holes are some of the most fascinating and mysterious objects in the universe. They are characterized by their strong gravitational pull, which can trap anything that enters their event horizon, and by their properties of mass, spin, and charge. Black holes play a crucial role in the formation and evolution of galaxies, and they provide a unique testing ground for our understanding of general relativity and the nature of space-time.

While black holes have been extensively studied and observed, there are still many unanswered questions and challenges, such as the nature of the singularity at the center of a black hole and the problem of information loss. Further research and study of black holes will undoubtedly lead to new discoveries and insights into the behavior of matter and energy in the universe.

27

Stellar Evolution and Black Hole Formation

Stellar evolution is the process by which a star changes over the course of its lifetime, eventually leading to its death and potentially the formation of a black hole. In this essay, we will explore the stages of stellar evolution and the conditions required for the formation of a black hole.

Birth of a Star

Stars are born from clouds of gas and dust, called nebulae, which are found throughout the galaxy. The process of star formation begins when a region of a nebula becomes dense enough to collapse under its own gravity. As the gas and dust in the region collapse, they begin to heat up due to the conversion of gravitational potential energy into thermal energy.

At some point, the temperature and pressure in the core of the collapsing gas become high enough to ignite nuclear fusion, the process by which atomic nuclei combine to form heavier elements and release energy. This marks the birth of a star and the beginning of the main sequence phase of its life.

Main Sequence Phase

The main sequence phase is the longest phase in a star's life, lasting tens of millions to billions of years depending on the star's mass. During this phase, the star fuses hydrogen into helium in its core, releasing energy in the process. The energy produced by nuclear fusion balances the inward pull of gravity, preventing the star from collapsing further.

The main sequence phase is characterized by a stable balance between the inward pull of gravity and the outward pressure due to nuclear fusion. The temperature and pressure in the core of the star remain relatively constant, and the star shines steadily for the duration of the phase.

Red Giant Phase

As the star exhausts the hydrogen in its core, the core begins to contract and heat up, causing the outer layers of the star to expand and cool. This marks the beginning of the red giant phase.

During the red giant phase, the star fuses helium into heavier elements, such as carbon and oxygen, in its core. The increased temperature and pressure in the core cause the outer layers of the star to expand, making the star much larger and cooler than it was during the main sequence phase.

Planetary Nebula and White Dwarf Phase

After the red giant phase, the outer layers of the star are expelled in a process called a planetary nebula. This leaves behind a hot, dense core known as a white dwarf.

White dwarfs are supported by electron degeneracy pressure, which arises from the exclusion principle of quantum mechanics. This pressure prevents the electrons in the core from occupying the same energy state, and it supports the star against further collapse.

Black Hole Formation

For stars that are more massive than about three times the mass of the sun, the story does not end with the formation of a white dwarf. In these stars, the core eventually becomes hot enough to fuse heavier elements, such as iron, but fusion of iron does not release energy, and therefore cannot support the star against gravitational collapse.

As the core collapses, it becomes denser and hotter, eventually reaching temperatures and densities so high that the atomic nuclei are crushed together and the protons and electrons merge to form neutrons. This marks the formation of a neutron star.

For stars that are even more massive, however, the gravitational collapse does not stop at a neutron star. Instead, the core continues to collapse until it becomes a singularity, a point of zero volume and infinite density, surrounded by an event horizon from which nothing can escape. This marks the formation of a black hole.

The conditions required for the formation of a black hole are a core mass greater than about three times the mass of the sun and a compactness greater than the Schwarzschild radius, a measure of the size of the event horizon for a given mass. These conditions can be achieved in massive stars during the final stages of their evolution.

Stellar evolution is a fascinating process that leads to the formation of stars, the production of heavy elements, and potentially the formation of black holes. The stages of stellar evolution, including the main sequence phase, red giant phase, and white dwarf phase, are well-understood and have been observed in many stars throughout the galaxy.

However, the formation of black holes remains a mysterious and fascinating area of research. The conditions required for black hole formation, including a core mass greater than about three times the mass of the sun and a compactness greater than the Schwarzschild radius, have been well-established, but the exact nature of the singularity at the center of a black hole remains unknown.

Further research into the formation and behavior of black holes will undoubtedly lead to new discoveries and insights into the nature of the universe and the behavior of matter and energy at the most fundamental levels.

28

The Anatomy of a Black Hole

A black hole is one of the most mysterious and enigmatic objects in the universe. It is a region in space where gravity is so strong that nothing, not even light, can escape its grasp. In this essay, we will explore the anatomy of a black hole, including its event horizon, singularity, and properties such as mass, spin, and charge.

Event Horizon

The event horizon is the boundary around a black hole beyond which nothing can escape. It is defined as the distance from the center of the black hole where the escape velocity is equal to the speed of light. Anything that crosses the event horizon is trapped within the gravitational pull of the black hole and cannot escape.

The size of the event horizon depends on the mass of the black hole. The larger the mass of the black hole, the larger the event horizon. For example, a black hole with a mass of one solar mass has an event horizon with a radius of about 3 kilometers, while a black hole with a mass of 10 solar masses has an event horizon with a radius of about 30 kilometers.

The event horizon is also the point at which time appears to stop for an outside observer. This is because the gravitational pull of the black hole is so strong that time slows down as you get closer to the event horizon. As you approach the event horizon, time appears to slow down more and more until it appears to stop altogether at the event horizon.

Singularity

The singularity is the point at the center of a black hole where the gravitational pull becomes infinite and space-time is infinitely curved. It is a point of zero volume and infinite density, and our current understanding of physics breaks down at this point.

The singularity is surrounded by the event horizon, which marks the boundary beyond which nothing can escape the black hole's gravitational pull. Anything that crosses the event horizon is inevitably pulled towards the singularity, where it is crushed into an infinitely small point.

The properties of the singularity are not well understood, as our current understanding of physics does not apply at such extreme conditions. It is not clear what happens to matter that falls into a black hole or how the singularity behaves.

Mass, Spin, and Charge

Black holes are characterized by their mass, spin, and charge. The mass of a black hole is the amount of matter that has collapsed into the singularity, and it determines the strength of the black hole's gravitational pull. The more massive a black hole is, the stronger its gravitational pull.

The spin of a black hole is a measure of how fast it is rotating. The faster a black hole is spinning, the more distorted space-time becomes near the black hole. The spin of a black hole can be measured indirectly by observing the effects of its gravitational pull on nearby matter, such as stars and gas clouds.

Black holes can also have an electric charge, which is a measure of the distribution of electric charge within the black hole. However, most black holes are believed to be neutral, meaning they have no electric charge.

Observing Black Holes

Black holes themselves cannot be directly observed, as they do not emit any light or radiation. However, the effects of black holes can be observed indirectly through their interactions with other matter and their gravitational influence on nearby objects.

One way that black holes can be detected is through their effects on nearby stars. If a star is orbiting around a black hole, its orbit can be measured and used to infer the mass of the black hole.

Another way that black holes can be detected is through the emission of radiation from matter falling into the black hole. As matter falls towards the black hole, it heats up and emits X-rays and other forms of radiation. This radiation can be detected by telescopes and used to infer the presence of a black hole.

the anatomy of a black hole is defined by its event horizon, singularity, and properties such as mass, spin, and charge. The event horizon marks the boundary beyond which nothing can escape the black hole's gravitational pull, while the singularity is the point at the center of the black hole where the gravitational pull becomes infinite and space-time is infinitely curved. The properties of the singularity are not well understood, as our current understanding of physics breaks down at this point.

Black holes are characterized by their mass, spin, and charge, which determine the strength of their gravitational pull and the effects of their interactions with nearby matter. Observing black holes directly is not possible, but their effects on nearby matter and their gravitational influence can be observed indirectly through telescopes and other instruments.

Black holes are some of the most mysterious and enigmatic objects in the universe, and their study provides valuable insights into the behavior of matter and energy under extreme conditions. While there is still much to learn about the nature of black holes, continued research and study will undoubtedly lead to new discoveries and a better understanding of the universe in which we live.

29

The Event Horizon and the Singularity

The concept of a black hole is one of the most mysterious and enigmatic in modern physics. At the heart of this concept lies two key components - the event horizon and the singularity. In this essay, we will explore the meaning and significance of these two components of black holes.

The Event Horizon

The event horizon is a boundary around a black hole beyond which nothing, not even light, can escape. It is defined as the distance from the center of the black hole where the escape velocity is equal to the speed of light. Anything that crosses the event horizon is trapped within the gravitational pull of the black hole and cannot escape.

The size of the event horizon depends on the mass of the black hole. The larger the mass of the black hole, the larger the event horizon. For example, a black hole with a mass of one solar mass has an event horizon with a radius of about 3 kilometers, while a black hole with a mass of 10 solar masses has an event horizon with a radius of about 30 kilometers.

The event horizon is also the point at which time appears to stop for an outside observer. This is because the gravitational pull of the black hole is so strong that time slows down as you get closer to the event horizon. As you approach the event horizon, time appears to slow down more and more until it appears to stop altogether at the event horizon.

The concept of the event horizon was first introduced by the physicist John

Michell in 1783. Michell proposed the existence of "dark stars," which were so massive that their gravity would be strong enough to prevent light from escaping. However, it wasn't until the early 20th century that the concept of the event horizon was fully understood and incorporated into the theory of general relativity.

The Singularity

The singularity is the point at the center of a black hole where the gravitational pull becomes infinite and space-time is infinitely curved. It is a point of zero volume and infinite density, and our current understanding of physics breaks down at this point.

The singularity is surrounded by the event horizon, which marks the boundary beyond which nothing can escape the black hole's gravitational pull. Anything that crosses the event horizon is inevitably pulled towards the singularity, where it is crushed into an infinitely small point.

The properties of the singularity are not well understood, as our current understanding of physics does not apply at such extreme conditions. It is not clear what happens to matter that falls into a black hole or how the singularity behaves.

The concept of the singularity was first proposed by the physicist and mathematician John Wheeler in the mid-20th century. Wheeler coined the term "black hole" and helped to develop the modern understanding of black holes as regions of space where the gravitational pull is so strong that nothing can escape.

The Properties of Black Holes

Black holes are characterized by their mass, spin, and charge. The mass of a black hole is the amount of matter that has collapsed into the singularity, and it determines the strength of the black hole's gravitational pull. The more massive a black hole is, the stronger its gravitational pull.

The spin of a black hole is a measure of how fast it is rotating. The faster a black hole is spinning, the more distorted space-time becomes near the black hole. The spin of a black hole can be measured indirectly by observing the effects of its gravitational pull on nearby matter, such as stars and gas clouds.

Black holes can also have an electric charge, which is a measure of the distribution of electric charge within the black hole. However, most black holes are believed to be neutral, meaning they have no electric charge.

Observing Black Holes

Black holes themselves cannot be directly observed, as they do not emit any light or radiation. However, the effects of black holes can be observed indirectly through their interactions with other matter and their gravitational influence on nearby objects.

One way that black holes can be detected is through their effects on nearby stars. If a star is orbiting around a black hole, its orbit can be measured and used to infer the mass of the black hole.

Another way that black holes can be detected is through the emission of radiation from matter falling into the black hole. As matter falls towards the black hole, it heats up and emits X-rays and other forms of radiation. This radiation can be detected by telescopes and used to infer the presence of a black hole.

The study of black holes provides valuable insights into the behavior of matter and energy under extreme conditions. While there is still much to learn about the nature of black holes, continued research and study will undoubtedly lead to new discoveries and a better understanding of the universe in which we live.

In conclusion, the event horizon and the singularity are the two key components of black holes. The event horizon marks the boundary beyond which nothing can escape the black hole's gravitational pull, while the singularity is the point at the center of the black hole where the gravitational pull becomes infinite and space-time is infinitely curved. The properties of black holes, including their mass, spin, and charge, determine the strength of their gravitational pull and the effects of their interactions with nearby matter.

While black holes themselves cannot be directly observed, their effects on nearby matter and their gravitational influence can be observed indirectly through telescopes and other instruments. The study of black holes provides valuable insights into the behavior of matter and energy under extreme conditions, and continued research and study will undoubtedly lead to new discoveries and a better understanding of the universe in which we live.

30

Gravitational Waves and Black Hole Mergers

Gravitational waves are ripples in the fabric of space-time that are produced by the acceleration of massive objects. These waves were first predicted by Albert Einstein's theory of general relativity over a century ago, but it wasn't until the 21st century that they were detected directly. One of the most exciting and significant sources of gravitational waves is the merger of two black holes, which produces a burst of gravitational radiation that can be detected by sensitive instruments on Earth.

In this essay, we will explore the phenomenon of gravitational waves and the merger of black holes, including the physics behind these events, the technology used to detect them, and the implications of these discoveries for our understanding of the universe.

Gravitational Waves

Gravitational waves are a prediction of Einstein's theory of general relativity, which describes the behavior of space and time in the presence of matter and energy. According to general relativity, massive objects warp the fabric of space-time, creating a gravitational field that causes other objects to accelerate towards them.

When massive objects move or accelerate, they produce ripples in the fabric of space-time that propagate outwards at the speed of light. These ripples are known as gravitational waves, and they can be thought of as analogous to the waves that are produced when a stone is thrown into a pond.

Gravitational waves are extremely difficult to detect, as they are very weak and interact very weakly with matter. However, advances in technology have made it possible to detect gravitational waves using sensitive instruments such as laser interferometers.

The Detection of Gravitational Waves

The first direct detection of gravitational waves was made in 2015 by the Laser Interferometer Gravitational-Wave Observatory (LIGO) in the United States. LIGO consists of two detectors, located in Hanford, Washington, and Livingston, Louisiana, which are designed to measure tiny fluctuations in the distance between two mirrors caused by passing gravitational waves.

The detection of gravitational waves was a major milestone in physics, as it confirmed a key prediction of general relativity and opened up a new way of observing the universe. Since then, other gravitational wave observatories have been built around the world, including Virgo in Italy and KAGRA in Japan.

Black Hole Mergers

One of the most significant sources of gravitational waves is the merger of two black holes. When two black holes orbit around each other, they lose energy in the form of gravitational radiation, causing them to spiral towards each other at an accelerating rate.

As the two black holes approach each other, they begin to orbit around each other faster and faster, emitting stronger and stronger gravitational waves. Finally, the two black holes merge together to form a single, more massive black hole, releasing a burst of gravitational radiation that propagates outwards through space.

The detection of black hole mergers has been one of the most significant discoveries made using gravitational wave detectors. Black holes are some of the most enigmatic and mysterious objects in the universe, and their study provides valuable insights into the behavior of matter and energy under extreme conditions.

Implications for Physics

The detection of gravitational waves and black hole mergers has significant implications for our understanding of the universe and the laws of physics. For example, the observation of gravitational waves provides a new way of observing the behavior of matter and energy in the universe, allowing us to test our understanding of general relativity and the nature of space and time.

The detection of black hole mergers also provides valuable information about the properties of black holes, including their masses and spins. This information can be used to test theoretical models of black holes and to constrain the properties of dark matter and dark energy.

Finally, the detection of gravitational waves and black hole mergers opens up a new window on the universe, allowing us to observe phenomena that are invisible to traditional optical telescopes. This includes the study of the early universe, the behavior of matter and energy under extreme conditions, and the search for new sources of gravitational waves, such as neutron star mergers.

Future Directions

The field of gravitational wave astronomy is still in its early stages, and there is much to be learned about the nature of black holes, the behavior of matter and energy under extreme conditions, and the structure of the universe itself.

In the coming years, new gravitational wave observatories will be built, including the LIGO-India observatory and the Einstein Telescope in Europe. These observatories will be more sensitive and capable than current detectors, allowing us to detect even fainter signals and study black holes and other gravitational wave sources in greater detail.

There is also ongoing research into the development of new technologies for detecting gravitational waves, including the use of atom interferometry and space-based detectors such as LISA (the Laser Interferometer Space Antenna).

In conclusion, the detection of gravitational waves and black hole mergers represents a major breakthrough in our understanding of the universe and the laws of physics. Gravitational waves provide a new way of observing the behavior of matter and energy in the universe, allowing us to test our understanding of general relativity and the nature of space and time.

Black hole mergers are a significant source of gravitational waves and provide valuable information about the properties of black holes and the behavior of matter and energy under extreme conditions. The detection of gravitational waves and black hole mergers opens up a new window on the universe, allowing us to observe phenomena that are invisible to traditional optical telescopes.

The field of gravitational wave astronomy is still in its early stages, and there is much to be learned about the nature of black holes, the behavior of matter and energy under extreme conditions, and the structure of the universe itself. Continued research and study in this area will undoubtedly lead to new discoveries and a better understanding of the universe in which we live.

VII

Black Holes Ain't So Black

31

Stephen Hawking and Black Hole Radiation

Stephen Hawking was a renowned physicist and cosmologist who made significant contributions to our understanding of the universe. One of his most famous contributions was the discovery of black hole radiation, which revolutionized our understanding of black holes and the behavior of matter and energy under extreme conditions.

In this essay, we will explore the discovery of black hole radiation by Stephen Hawking, including the physics behind this phenomenon and its implications for our understanding of black holes and the universe.

The Discovery of Black Hole Radiation

In the 1970s, Stephen Hawking began to investigate the behavior of black holes, which are some of the most enigmatic and mysterious objects in the universe. Black holes are regions of space where the gravitational pull is so strong that nothing, not even light, can escape. As a result, they are invisible to traditional telescopes and can only be detected through their effects on nearby matter and radiation.

Hawking's research led him to an astonishing conclusion - that black holes were not completely black after all. Instead, he proposed that black holes emitted radiation due to quantum effects near the event horizon, the boundary around the black hole beyond which nothing can escape.

According to Hawking's theory, pairs of virtual particles and anti-particles are constantly being created near the event horizon of a black hole. Normally, these pairs quickly annihilate each other, and their energy is lost. However, if one member of the pair falls into the black hole, while the other escapes, then it is possible for the black hole to lose energy through the process of Hawking radiation.

The energy loss caused by Hawking radiation causes the black hole to gradually shrink and eventually evaporate completely. This process, which takes an incredibly long time for all but the smallest black holes, is known as black hole evaporation.

Implications for Black Hole Physics

The discovery of black hole radiation had significant implications for our understanding of black holes and the behavior of matter and energy under extreme conditions. Prior to Hawking's discovery, black holes were thought to be completely black, with nothing able to escape their gravitational pull.

The existence of Hawking radiation meant that black holes could lose energy and eventually evaporate completely. This had significant implications for our understanding of the fate of black holes and the ultimate structure of the universe.

Furthermore, Hawking's discovery of black hole radiation provided a new way of studying the properties of black holes, such as their mass, spin, and charge. The observation of Hawking radiation could provide valuable information about the nature of black holes and the behavior of matter and energy near their event horizons.

Challenges and Controversies

Despite the significance of Hawking's discovery, the concept of black hole radiation remains controversial and has been the subject of much debate in the physics community. Some physicists have challenged the theory, arguing that it violates the laws of thermodynamics or that it is incompatible with other aspects of quantum mechanics.

Others have proposed alternative theories of black hole evaporation, such as the idea that black holes emit particles known as "firewalls" near their event horizons. These alternative theories remain the subject of ongoing research and debate in the physics community.

Stephen Hawking's discovery of black hole radiation was a significant breakthrough in our understanding of black holes and the behavior of matter and energy under extreme conditions. The concept of Hawking radiation provided a new way of studying the properties of black holes and had significant implications for our understanding of the ultimate fate of these enigmatic objects.

However, the concept of black hole radiation remains controversial and has been the subject of much debate in the physics community. Ongoing research and study will undoubtedly lead to new discoveries and a better understanding of the universe in which we live.

Despite the controversies surrounding black hole radiation, it remains one of the most fascinating and important discoveries in the field of physics. The concept of Hawking radiation has inspired new avenues of research and inquiry, as well as new ways of understanding the behavior of matter and energy under extreme conditions.

Moreover, Hawking's contributions to physics and cosmology extend far beyond his discovery of black hole radiation. His work on the nature of time, the origin of the universe, and the relationship between quantum mechanics and gravity has had a profound impact on our understanding of the universe and our place within it.

In summary, the discovery of black hole radiation by Stephen Hawking was a groundbreaking achievement in the field of physics. It challenged our assumptions about the behavior of black holes and opened up new avenues of research and inquiry. While the concept of black hole radiation remains controversial and subject to ongoing debate, its implications for our understanding of the universe and the laws of physics are undeniable.

32

The Information Paradox

The information paradox is a long-standing problem in the field of black hole physics that was first proposed by Stephen Hawking in the 1970s. The paradox arises from the fact that black holes seem to violate the laws of quantum mechanics, which describe the behavior of matter and energy at the subatomic level.

In this essay, we will explore the information paradox and its implications for our understanding of black holes and the laws of physics. We will also discuss some proposed solutions to the paradox and ongoing research in this area.

The Paradox

According to the laws of quantum mechanics, information cannot be destroyed. When two quantum particles interact, they become entangled, meaning that the state of one particle is dependent on the state of the other particle. This entanglement persists even if the particles are separated by large distances, and it cannot be undone without destroying the particles themselves.

In the case of black holes, the paradox arises because black holes are thought to destroy information. According to classical physics, anything that falls into a black hole is lost forever, and there is no way for any information about the object to escape.

However, according to quantum mechanics, information cannot be destroyed. If information is lost when an object falls into a black hole, then this violates the laws of quantum mechanics, which state that information must be conserved.

Solutions to the Paradox

There have been many proposed solutions to the information paradox over the years, ranging from modifications to classical physics to more radical ideas such as the holographic principle.

One proposed solution is known as the "firewall" hypothesis. According to this hypothesis, the information that falls into a black hole is stored on the event horizon, which acts as a firewall that destroys anything that tries to cross it.

Another proposed solution is the idea of black hole complementarity. According to this idea, there are two different ways of describing a black hole - one from the outside observer's point of view and one from the point of view of an observer who falls into the black hole. From the outside observer's point of view, information is lost when an object falls into a black hole, but from the perspective of an observer falling into the black hole, the information is preserved.

The holographic principle is another proposed solution to the information paradox. This principle states that all the information contained within a three-dimensional space can be represented by a two-dimensional hologram. According to this idea, the information that falls into a black hole is stored on its two-dimensional event horizon, rather than in its three-dimensional interior.

Ongoing Research

Despite the many proposed solutions to the information paradox, the problem remains unresolved, and ongoing research is being conducted in this area. One promising approach is the study of quantum entanglement, which may provide insights into how information is preserved in black holes.

Another area of research is the study of black hole evaporation, which is thought to be the mechanism by which black holes lose mass and eventually disappear. The study of black hole evaporation has led to new insights into the behavior of black holes and their interactions with matter and energy.

In conclusion, the information paradox is a long-standing problem in the field of black hole physics that arises from the apparent violation of the laws of quantum mechanics by black holes. There have been many proposed solutions to the paradox over the years, ranging from modifications to classical physics to more radical ideas such as the holographic principle.

Despite the many proposed solutions, the problem remains unresolved, and ongoing research is being conducted in this area. The study of black holes and their interactions with matter and energy is a fascinating and important area of research that has profound implications for our understanding of the universe and the laws of physics.

33

Black Hole Entropy and the Holographic Principle

The concept of black hole entropy and the holographic principle are important areas of research in the field of theoretical physics, with profound implications for our understanding of black holes, the nature of information, and the laws of the universe.

In this essay, we will explore the concept of black hole entropy and the holographic principle, including their origins, implications, and ongoing research in this area.

Black Hole Entropy

Entropy is a measure of the disorder or randomness of a system. In the case of black holes, entropy is a measure of the number of different ways in which matter and energy can be arranged within the black hole. According to classical physics, the entropy of a black hole should be zero, since all of the matter and energy within the black hole is thought to be compressed into a single point.

However, in the 1970s, Stephen Hawking proposed that black holes actually do have entropy, due to the fact that they emit thermal radiation (known as Hawking radiation) and gradually lose mass over time. Hawking radiation is caused by the creation of particle-antiparticle pairs near the event horizon of the black hole, with one particle falling into the black hole and the other escaping as radiation. The emission of radiation causes the black hole to gradually lose mass and, according to the laws of thermodynamics, also causes an increase in the entropy of the black hole.

The Holographic Principle

The holographic principle is a concept that has its origins in the study of black hole physics, but which has since been applied to other areas of physics, including the study of the early universe and the behavior of matter and energy at the subatomic level.

The holographic principle states that all of the information contained within a three-dimensional space can be represented by a two-dimensional hologram. This means that the entire contents of a volume of space, including its matter and energy, can be described by information contained on its surface.

In the case of black holes, the holographic principle suggests that all of the information contained within a black hole can be represented by the information contained on its event horizon. This means that the contents of a black hole, including its entropy, can be described by information contained on its surface, rather than within its interior.

Implications for Physics

The concept of black hole entropy and the holographic principle have significant implications for our understanding of black holes, the nature of information, and the laws of the universe.

For example, the existence of black hole entropy challenges our understanding of the relationship between information and the laws of thermodynamics. The second law of thermodynamics states that the entropy of a closed system must always increase over time, but the existence of black hole entropy suggests that this law may not be absolute.

Furthermore, the holographic principle has led to new insights into the nature of space and time, and the relationship between quantum mechanics and gravity. The holographic principle suggests that space and time may be emergent properties of the universe, rather than fundamental aspects of reality, and that the laws of quantum mechanics may be able to describe the behavior of matter and energy in the universe without the need for a separate theory of gravity.

Ongoing Research

Despite the many advances that have been made in the study of black hole entropy and the holographic principle, many questions remain unanswered. Ongoing research is being conducted in this area, with the aim of better understanding the nature of black holes, the behavior of matter and energy in extreme conditions, and the laws of the universe.

For example, researchers are studying the relationship between black hole entropy and the holographic principle, with the aim of better understanding how information is stored and processed within black holes. Other researchers are studying the implications of the holographic principle for the nature of space and time, and the relationship between quantum mechanics and gravity.

the concept of black hole entropy and the holographic principle are important areas of research in the field of theoretical physics. Black hole entropy challenges our understanding of the relationship between information and the laws of thermodynamics, while the holographic principle has led to new insights into the nature of space and time, and the relationship between quantum mechanics and gravity.

Ongoing research in this area has the potential to unlock new insights into the behavior of matter and energy in extreme conditions, the nature of black holes, and the laws of the universe. The study of black hole entropy and the holographic principle is a fascinating and important area of research that has the potential to revolutionize our understanding of the universe and the laws of physics.

34

Black Holes as Quantum Computers

The concept of black holes as quantum computers is a relatively new area of research that has the potential to revolutionize our understanding of both black holes and quantum computing. In this chapter, we will explore the concept of black holes as quantum computers, including its origins, implications, and ongoing research in this area.

Quantum Computing

Quantum computing is a type of computing that is based on the principles of quantum mechanics, rather than classical mechanics. Unlike classical computing, which uses bits to store and process information, quantum computing uses quantum bits (or qubits) that can exist in multiple states simultaneously. This allows quantum computers to perform certain calculations much faster than classical computers.

One of the most promising applications of quantum computing is in the field of cryptography, where it is thought that quantum computers could be used to break many of the currently-used cryptographic codes. However, quantum computing is still in its early stages of development, and there are many technical challenges that must be overcome before practical quantum computers can be built.

Black Holes as Quantum Computers

The idea of black holes as quantum computers was first proposed by Seth Lloyd in 2002. According to this idea, the event horizon of a black hole acts as a quantum computer, with the information stored on the event horizon being processed and computed using the principles of quantum mechanics.

The event horizon of a black hole is the point of no return, beyond which nothing can escape. According to Lloyd's proposal, the event horizon of a black hole can be thought of as a two-dimensional surface, which can be used to store and process information. This information is thought to be stored in the form of qubits, which are entangled with the matter and energy that falls into the black hole.

Implications for Physics

The concept of black holes as quantum computers has significant implications for our understanding of both black holes and quantum computing. For example, if black holes are indeed quantum computers, then they may be able to perform certain calculations much faster than classical computers, due to their ability to process information using the principles of quantum mechanics.

Furthermore, the concept of black holes as quantum computers suggests that there may be a deeper relationship between black holes and quantum mechanics than previously thought. This relationship could potentially lead to new insights into the behavior of matter and energy in extreme conditions, and the nature of the universe itself.

Ongoing Research

Despite the potential implications of the concept of black holes as quantum computers, there is still much research that needs to be done in this area. One of the biggest challenges is the fact that the information that falls into a black hole is thought to be irretrievable, due to the fact that it is compressed and destroyed by the immense gravitational forces within the black hole.

However, there are some proposed solutions to this problem, such as the idea of black hole complementarity, which suggests that the information that falls into a black hole is not actually destroyed, but is instead encoded in the radiation that is emitted by the black hole.

Another area of ongoing research is the study of the relationship between black holes and quantum mechanics, and the potential implications of this relationship for our understanding of the universe and the laws of physics.

In conclusion, the concept of black holes as quantum computers is a fascinating area of research that has the potential to revolutionize our understanding of both black holes and quantum computing. While there are many technical challenges that must be overcome before practical black hole quantum computers can be built, the implications of this concept for our understanding of the universe and the laws of physics are profound. Ongoing research in this area has the potential to unlock new insights into the behavior of matter and energy in extreme conditions, and the nature of the universe itself.

One of the key equations used in the study of black holes as quantum computers is the Bekenstein-Hawking formula, which relates the entropy of a black hole to its surface area:

$S = kA/4l\_P^2$

In this formula, S represents the entropy of the black hole, A represents its surface area, k is the Boltzmann constant, and l_P is the Planck length.

The Bekenstein-Hawking formula is important in the study of black holes as quantum computers because it suggests that the information stored on the event horizon of a black hole is directly related to the entropy of the black hole. This implies that the information stored on the event horizon can be thought of as a form of quantum information, which can be processed and computed using the principles of quantum mechanics.

Another important equation used in the study of black holes as quantum computers is the Wheeler-DeWitt equation, which describes the wave function of the universe. This equation is important because it suggests that the behavior of matter and energy in the universe is closely related to the behavior of matter and energy within black holes.

The study of black holes as quantum computers is still in its early stages, and there is much that is still unknown about the behavior of matter and energy within black holes. However, ongoing research in this area has the potential to unlock new insights into the nature of the universe and the laws of physics.

35

The Firewall Paradox

The Firewall Paradox is a fascinating concept in the realm of theoretical physics, stirring up debates and discussions among scientists for years. It presents a challenge to the fundamental principles of quantum mechanics and general relativity when attempting to understand the nature of black holes. This paradox, also known as the black hole information paradox, is a thought experiment that tries to reconcile seemingly contradictory aspects of black hole physics. In this essay, we will delve into the origins of the Firewall Paradox, its implications on our understanding of the universe, and the ongoing debates among scientists attempting to resolve this enigma.

A Brief History of Black Holes

The concept of black holes dates back to the 18th century when English scientist John Michell and French scientist Pierre-Simon Laplace independently proposed the idea of a "dark star," whose gravity would be so strong that not even light could escape it. However, it wasn't until the 20th century, when Albert Einstein formulated his theory of general relativity, that the modern understanding of black holes began to take shape.

In 1915, Einstein published his groundbreaking theory, which provided a new way of understanding gravity as a curvature of spacetime caused by mass. A few years later, in 1916, the German physicist Karl Schwarzschild found a solution to Einstein's equations that described the spacetime geometry around a spherical, non-rotating mass. This solution laid the foundation for the modern concept of a black hole, characterized by an event horizon, a boundary from which nothing can escape, including light.

The Firewall Paradox: Origins and Formulation

In the late 20th century, the physicist Stephen Hawking made a remarkable discovery that seemed to challenge the classical understanding of black holes. Hawking's calculations showed that black holes could emit radiation, now known as Hawking radiation, due to quantum effects occurring near the event horizon. This radiation would cause the black hole to lose mass over time, eventually leading to its evaporation.

The idea of black hole evaporation created a significant problem in the world of physics – the information loss paradox. According to the fundamental principles of quantum mechanics, information must be conserved. However, if a black hole evaporates and disappears, taking all the information about the matter it consumed with it, this principle would be violated.

In 2012, a group of physicists, including Joseph Polchinski, Ahmed Almheiri, Donald Marolf, and James Sully, proposed a solution to the information paradox called the Firewall Paradox. Their solution was based on the concept of entanglement, a fundamental aspect of quantum mechanics that links particles in such a way that the properties of one are intrinsically connected to the properties of the other, regardless of the distance between them.

The Firewall Paradox arises from the following considerations:

1. If an infalling particle becomes entangled with an outgoing Hawking radiation particle, then the information about the infalling particle can, in principle, escape the black hole and be preserved.
2. However, for the information to be conserved outside the black hole, the outgoing Hawking radiation particles must also be entangled with each other.
3. According to the "monogamy of entanglement" principle, a particle cannot be fully entangled with two other particles simultaneously.

These considerations lead to a paradoxical conclusion. If the infalling particle is entangled with the outgoing Hawking radiation particle to preserve information, it violates the monogamy of entanglement, as the Hawking radiation particles must also be entangled with each other. This conflict gives rise to the concept of a "firewall," a hypothetical, highly energetic barrier just inside the event horizon, which severs the entanglement between the infalling and outgoing particles, thus preserving the monogamy of entanglement. However, the existence of such a firewall would contradict the equivalence principle, a cornerstone of general relativity, which states that a free-falling observer should not experience any unusual effects while crossing the event horizon of a black hole.

Implications and Ongoing Debates

The Firewall Paradox has generated intense debate among physicists, as it highlights a fundamental inconsistency between quantum mechanics and general relativity. Resolving the paradox would require modifying one of these theories or finding a way to reconcile the seemingly contradictory principles. Several potential solutions and alternative explanations have been proposed, each with its own merits and challenges.

1. The ER=EPR Conjecture: In 2013, physicist Juan Maldacena proposed a conjecture that connects wormholes (Einstein-Rosen bridges) to quantum entanglement (Einstein-Podolsky-Rosen paradox). This conjecture suggests that entangled particles are connected by wormholes, which could, in principle, allow information to escape from a black hole without violating the monogamy of entanglement. However, the ER=EPR conjecture is still speculative and not yet proven.

2. Soft Hair on Black Holes: In 2016, Stephen Hawking, along with Malcolm Perry and Andrew Strominger, proposed a potential resolution to the information paradox based on the concept of "soft hair" on black holes. They suggested that black holes could carry "soft hair," quantum excitations that could store the information about the infalling particles. This idea implies that the information is not lost but rather encoded on the event horizon, allowing it to be retrieved as the black hole evaporates. While intriguing, this proposal requires further development and rigorous testing.

3. Holographic Principle: Another approach to resolving the Firewall Paradox is based on the holographic principle, which posits that the information about a volume of space can be encoded on its boundary. In the context of black holes, the information about the infalling particles could be stored on the event horizon as a hologram. This principle has led to the development of the AdS/CFT correspondence, a powerful tool in theoretical physics that connects quantum gravity in Anti-de Sitter (AdS) space to conformal field theories (CFT) on its boundary. However, the holographic principle has yet to be fully established as a fundamental law of nature.

The Firewall Paradox continues to challenge our understanding of the universe, exposing deep-seated tensions between quantum mechanics and general relativity. Resolving this paradox could lead to new insights into the nature of black holes, the fabric of spacetime, and the ultimate theory that unifies gravity with the other fundamental forces. While many potential solutions have been proposed, none have yet gained consensus among physicists. The quest to unravel the mysteries of the Firewall Paradox and the enigmatic black holes remains an active and exciting area of research, promising to reshape our understanding of the cosmos.

VIII

The Origin and Fate of the Universe

36

Introduction to the Origin and Fate of the Universe

The origin and fate of the universe is one of the most fundamental questions in physics and cosmology. Scientists and astronomers have been studying the universe for centuries, seeking to understand how it began, how it has evolved over time, and what its ultimate fate might be.

In this essay, we will provide an introduction to the origin and fate of the universe, including the current scientific understanding of these topics and ongoing research in the field.

The Origin of the Universe

The current scientific understanding of the origin of the universe is based on the Big Bang theory. According to this theory, the universe began as a singularity, a point of infinite density and temperature, approximately 13.8 billion years ago.

Shortly after the Big Bang, the universe underwent a period of rapid expansion known as inflation. During this time, the universe expanded faster than the speed of light, increasing in size by a factor of at least 10^{26}.

As the universe cooled, matter and energy began to form, eventually leading to the formation of atoms and the first galaxies. The earliest galaxies formed approximately 200-300 million years after the Big Bang.

The Fate of the Universe

The ultimate fate of the universe is still a topic of active research and debate. There are several different possibilities, depending on the parameters of the universe and the behavior of its constituents.

One possibility is that the universe will continue to expand indefinitely, eventually reaching a state of maximum entropy known as the heat death of the universe. In this scenario, all matter and energy will be evenly distributed throughout the universe, with no more thermodynamic free energy available to do work.

Another possibility is that the universe will eventually collapse in on itself in a Big Crunch, leading to another singularity similar to the one that existed at the beginning of the universe.

A third possibility is that the universe will continue to expand, but at an accelerating rate due to the presence of dark energy. This scenario is known as the Big Rip, and would result in the universe tearing itself apart as the expansion rate approaches infinity.

Ongoing Research

Despite the significant progress that has been made in our understanding of the origin and fate of the universe, there is still much that is unknown. Ongoing research in the field is focused on a variety of different topics, including the behavior of dark matter and dark energy, the nature of the early universe, and the behavior of black holes.

One area of active research is the study of cosmic microwave background radiation, which is the oldest light in the universe and provides clues about the conditions of the early universe. Another area of research is the study of gravitational waves, which were first detected in 2015 and provide a new way to study the behavior of matter and energy in the universe.

There is also ongoing research into the behavior of dark matter and dark energy, which are thought to make up approximately 95% of the mass-energy of the universe. The nature of these mysterious substances is still not fully understood, but their behavior has important implications for the ultimate fate of the universe.

There are several key equations used in the study of the origin and fate of the universe, including the following:

1. The Friedmann equations - These are a set of equations that describe the expansion of the universe based on Einstein's theory of general relativity. The equations relate the rate of expansion of the universe to the distribution of matter and energy within it. The Friedmann equations are:

$(H/H\_0)^2 = [(8\pi G)/3c^2]\rho - (kc^2)/R^2 + \Lambda/3$

where H is the Hubble parameter (the rate of expansion of the universe), H_0 is the present value of the Hubble parameter, G is the gravitational constant, c is the speed of light, ρ is the density of matter and energy in the universe, k is the curvature of space-time, R is the scale factor (the size of the universe at a given time), and Λ is the cosmological constant.

1. The equation of state of dark energy - This equation describes the relationship between the pressure and density of dark energy, a mysterious substance that is thought to be responsible for the accelerating expansion of the universe. The equation of state is:

$P = w\rho c^2$

where P is the pressure of dark energy, ρ is its density, c is the speed of light, and w is a parameter that characterizes the behavior of dark energy.

1. The equation of state of matter - This equation describes the relationship between the pressure and density of matter in the universe. The equation of state is:

$P = \gamma\rho c^2$

where P is the pressure of matter, ρ is its density, c is the speed of light, and γ is a parameter that characterizes the behavior of matter.

1. The energy-momentum tensor - This tensor describes the distribution of matter and energy in the universe and is used in Einstein's field equations to describe the curvature of space-time. The energy-momentum

tensor is:

$$T^{\wedge}(\mu\nu) = (\rho c^{\wedge}2 + P)u^{\wedge}(\mu)u^{\wedge}(\nu) - Pg^{\wedge}(\mu\nu)$$

where $T^{\wedge}(\mu\nu)$ is the energy-momentum tensor, ρ is the density of matter and energy, P is the pressure, $u^{\wedge}(\mu)$ is the four-velocity of matter and energy, and $g^{\wedge}(\mu\nu)$ is the metric tensor.

1. The Planck length - This is a fundamental constant of nature that is used in many equations in the study of the origin and fate of the universe. The Planck length is:

$$l\_P = \sqrt{(\hbar G/c^{\wedge}3)}$$

where ℏ is the reduced Planck constant, G is the gravitational constant, and c is the speed of light.

These equations, and others, are used to describe the behavior of matter and energy in the universe, the expansion of the universe, and the nature of dark matter and dark energy. They provide a framework for understanding the origin and fate of the universe and are important tools in ongoing research in the field.

The origin and fate of the universe are some of the most fundamental questions in physics and cosmology. The current scientific understanding of the origin of the universe is based on the Big Bang theory, while the ultimate fate of the universe is still a topic of active research and debate.

Ongoing research in the field is focused on a variety of different topics, including the behavior of dark matter and dark energy, the nature of the early universe, and the behavior of black holes. Despite the significant progress that has been made, there is still much that is unknown about the origin and fate of the universe, and ongoing research in this area has the potential to unlock new insights into the nature of the universe and the laws of physics.

37

The Inflationary Universe

The Inflationary Universe theory is a model of the early universe that seeks to explain the uniformity and flatness of the universe at large scales. The theory proposes that the universe underwent a brief period of extremely rapid expansion, known as inflation, within the first fraction of a second after the Big Bang.

The Inflationary Universe theory was first proposed by cosmologist Alan Guth in 1981, and has since become one of the most widely accepted models of the early universe. In this essay, we will explore the Inflationary Universe theory, including its origins, its key concepts, and its impact on our understanding of the universe.

Origins of the Inflationary Universe Theory

The Inflationary Universe theory was developed as a response to two key problems in cosmology. The first problem was the horizon problem, which refers to the fact that different regions of the universe that are separated by vast distances appear to have the same temperature and properties, despite the fact that they could not have interacted with one another due to the finite speed of light.

The second problem was the flatness problem, which refers to the observation that the universe appears to be very flat on large scales, despite the fact that it should be curved due to the presence of matter and energy.

Alan Guth proposed that these problems could be solved by a period of inflationary expansion that occurred in the very early universe. During this period, the universe underwent an exponential expansion, increasing in size by a factor of at least $10^{\wedge}26$ within a fraction of a second. This rapid expansion would have caused regions of the universe that were initially in causal contact to become far apart from one another, explaining the horizon problem. It would also have smoothed out any curvature in the universe, leading to the observed flatness.

Key Concepts of the Inflationary Universe Theory

The Inflationary Universe theory is based on several key concepts, including the following:

1. The inflaton field - In the Inflationary Universe model, the rapid expansion of the universe is driven by a hypothetical scalar field known as the inflaton field. This field has a potential energy density that is responsible for the inflationary expansion.

2. Quantum fluctuations - During the inflationary period, quantum fluctuations in the inflaton field would have been amplified and stretched to cosmic scales. These fluctuations are thought to be responsible for the large-scale structure of the universe, including the distribution of galaxies and clusters.

3. Reheating - After the inflationary period ended, the inflaton field decayed into particles, leading to a period of reheating in which the universe was filled with radiation and matter.

Impact of the Inflationary Universe Theory

The Inflationary Universe theory has had a significant impact on our understanding of the early universe and the origin of cosmic structure. It has helped to explain the uniformity and flatness of the universe on large scales, as well as the origin of the large-scale structure of the universe.

The theory has also made several predictions that have been confirmed by observations. For example, it predicts that the universe should be filled with a background of gravitational waves produced during the inflationary period. These gravitational waves have been detected by the BICEP and Planck experiments.

Furthermore, the Inflationary Universe theory has had a profound impact on our understanding of the nature of the universe as a whole. It suggests that the universe may be much larger than we previously thought, and may be just one of many universes in a multiverse.

The Inflationary Universe theory is a model of the early universe that seeks to explain the uniformity and flatness of the universe at large scales. It proposes that the universe underwent a period of rapid expansion, known as inflation, within the first fraction of a second after the Big Bang. This theory has had a The Inflationary Universe theory involves several key equations that describe the dynamics of the universe during the period of inflation. Some of the most important equations are:

1. The energy density of the inflaton field - The energy density of the inflaton field, which drives the inflationary expansion, can be described by the following equation:

$\rho = (1/2)\dot{\varphi}^2 + V(\varphi)$

where ρ is the energy density, $\dot{\varphi}$ is the time derivative of the inflaton field, and $V(\varphi)$ is the potential energy of the field.

1. The equation of motion for the inflaton field - The evolution of the inflaton field during the period of inflation is governed by the following equation of motion:

$\ddot{\varphi} + 3H\dot{\varphi} + V'(\varphi) = 0$

where H is the Hubble parameter (the rate of expansion of the universe), and $V'(\varphi)$ is the derivative of the potential energy with respect to the inflaton field.

1. The Friedmann equations during inflation - The evolution of the universe during the period of inflation can be described by the Friedmann

equations, which relate the rate of expansion of the universe to the distribution of matter and energy within it. During inflation, the Friedmann equations take the form:

$$H^2 = (8\pi G/3)\rho + (k/a^2) + (1/3)\Lambda$$

where H is the Hubble parameter, G is the gravitational constant, ρ is the energy density of matter and energy in the universe, k is the curvature of space-time, a is the scale factor (the size of the universe at a given time), and Λ is the cosmological constant.

1. The amplitude of density fluctuations - One of the most important predictions of the inflationary theory is that it should produce small, quantum fluctuations in the inflaton field that are stretched out to cosmic scales during inflation. These fluctuations are thought to be responsible for the large-scale structure of the universe. The amplitude of these fluctuations can be calculated using the following equation:

$$\delta\rho/\rho \approx H/2\pi^\varphi$$

where $\delta\rho/\rho$ is the amplitude of the density fluctuations, H is the Hubble parameter, and φ is the value of the inflaton field during inflation.

These equations, and others, are used to describe the behavior of the inflaton field, the expansion of the universe during inflation, and the generation of density fluctuations that give rise to the large-scale structure of the universe.

38

The Cosmic Censorship Hypothesis

The Cosmic Censorship Hypothesis is a conjecture in the field of general relativity that seeks to address the question of whether spacetime singularities can be observed from the outside of a black hole. First proposed by the British physicist Roger Penrose in 1969, this hypothesis plays a crucial role in our understanding of black holes and the fundamental structure of spacetime. In this essay, we will explore the origins, implications, and various forms of the Cosmic Censorship Hypothesis, along with the ongoing efforts to prove or disprove it.

Singularities and General Relativity

A singularity is a point in spacetime where the curvature becomes infinite, and the laws of physics as we know them break down. According to general relativity, singularities are expected to form at the center of black holes, as well as in the initial state of the universe, known as the Big Bang singularity.

The existence of singularities poses a challenge to our understanding of the universe, as it seems to imply that spacetime has a boundary beyond which we cannot predict or describe physical phenomena. This is at odds with the idea that the universe should be describable by a complete set of physical laws.

The Cosmic Censorship Hypothesis

In response to this conundrum, Roger Penrose proposed the Cosmic Censorship Hypothesis, which posits that all singularities in the universe must be hidden behind event horizons, effectively shielding them from observation by distant observers. In other words, nature "censors" the existence of singularities by enclosing them within black holes.

There are two primary forms of the Cosmic Censorship Hypothesis:

1. Weak Cosmic Censorship: This version of the hypothesis states that any singularity resulting from the gravitational collapse of a star must be hidden within a black hole's event horizon. This implies that singularities cannot be observed directly, but their existence can still influence the outside universe through the spacetime curvature surrounding the black hole.

2. Strong Cosmic Censorship: The strong version of the hypothesis goes further, stating that not only are singularities hidden behind event horizons, but also that the spacetime structure inside a black hole is stable and predictable, at least up to the point where the singularity forms. This would mean that the laws of physics remain valid and deterministic even within the black hole, except at the singularity itself.

Implications and Applications

The Cosmic Censorship Hypothesis has several important implications and applications in the fields of astrophysics and general relativity:

1. Black Hole Uniqueness Theorems: Assuming the validity of the Cosmic Censorship Hypothesis, it is possible to prove the black hole uniqueness theorems, which state that all stationary, asymptotically flat black holes can be fully characterized by just three parameters: mass, electric charge, and angular momentum. This is often referred to as the "no-hair theorem," suggesting that black holes have "no hair" or additional properties beyond these parameters.

2. Cosmic Censorship and the Penrose Process: The Penrose process is a theoretical mechanism by which energy can be extracted from a rotating black hole, potentially enabling the construction of advanced power sources in the far future. The validity of the Cosmic Censorship

Hypothesis is critical for the feasibility of the Penrose process, as it ensures that the extraction of energy does not lead to the exposure of a naked singularity.

The Cauchy Horizon Problem: The strong form of the Cosmic Censorship Hypothesis has implications for the so-called Cauchy horizon problem. The Cauchy horizon is a hypothetical boundary within a black hole, beyond which the initial conditions of the spacetime cannot uniquely determine its future evolution. If the strong cosmic censorship holds true, then the Cauchy horizon would be unstable, and the determinism of general relativity would be preserved up to the singularity itself.

Challenges and Counterexamples

The Cosmic Censorship Hypothesis, while widely accepted by many physicists, is not without its challenges and counterexamples. Over the years, researchers have discovered various scenarios in which singularities might be exposed to the outside universe, known as "naked singularities."

1. Rotating Black Holes: The Kerr solution to the Einstein field equations describes a rotating black hole, and it predicts that if a black hole's angular momentum exceeds a certain threshold, the event horizon disappears, leaving the singularity exposed. However, it is still unclear whether such a situation could arise through a realistic astrophysical process.

2. Gravitational Collapse: Some studies of gravitational collapse under specific conditions have suggested that it is possible to form naked singularities instead of black holes. These scenarios often involve fine-tuning of the initial conditions or highly contrived situations, which may not occur naturally in the universe.

3. Cosmic Censorship in Alternative Theories of Gravity: The validity of the Cosmic Censorship Hypothesis might be affected by modifications to general relativity, such as those proposed in some quantum gravity models. Investigations into these alternative theories have yielded mixed results, with some supporting cosmic censorship and others

suggesting the possibility of naked singularities.

Ongoing Research and Future Prospects

Despite its long history and numerous challenges, the Cosmic Censorship Hypothesis remains an open question in the field of theoretical physics. Ongoing research aims to explore its validity by studying gravitational collapse, the dynamics of black holes, and the interplay between general relativity and quantum mechanics.

Some of the key areas of investigation include:

1. Numerical Relativity: Powerful computational methods are being employed to simulate the gravitational collapse of massive objects and the subsequent formation of black holes or singularities. These simulations can provide valuable insights into the conditions under which cosmic censorship holds true or fails.

2. Quantum Gravity: The search for a consistent theory of quantum gravity, which combines general relativity with quantum mechanics, may help to shed light on the nature of singularities and the validity of the Cosmic Censorship Hypothesis. One of the leading candidates for such a theory is Loop Quantum Gravity, which has the potential to resolve singularities and provide a more complete description of the spacetime structure inside black holes.

3. Astrophysical Observations: Observations of black holes and other compact objects in the universe can provide crucial data to test the predictions of general relativity and the Cosmic Censorship Hypothesis. For example, the Event Horizon Telescope's groundbreaking image of the black hole at the center of the galaxy M87 has opened up new avenues for testing the no-hair theorem and the cosmic censorship conjecture.

The Cosmic Censorship Hypothesis is a conjecture, rather than a formal theorem with specific equations. However, it is closely related to the mathematics of general relativity and the study of black holes. Here, we provide some key equations and concepts relevant to the Cosmic Censorship

Hypothesis:

1. Einstein Field Equations:

The foundation of general relativity is the Einstein Field Equations (EFE), which relate the curvature of spacetime to the distribution of mass and energy:

$G_{\mu\nu} + \Lambda g_{\mu\nu} = (8\pi G/c^4)T_{\mu\nu}$

Here, $G_{\mu\nu}$ is the Einstein tensor, Λ is the cosmological constant, $g_{\mu\nu}$ is the metric tensor, G is the gravitational constant, c is the speed of light, and $T_{\mu\nu}$ is the stress-energy tensor.

1. Schwarzschild Metric:

The Schwarzschild solution is a spherically symmetric, non-rotating black hole solution to the Einstein Field Equations:

$ds^2 = -(1 - 2GM/c^2r)c^2dt^2 + (1 - 2GM/c^2r)^{-1}dr^2 + r^2(d\theta^2 + \sin^2\theta \, d\varphi^2)$

Here, ds^2 is the spacetime interval, M is the mass of the black hole, and (r, θ, φ) are the usual spherical coordinates.

1. Kerr Metric:

The Kerr solution is a more general solution to the EFE, representing a rotating black hole:

$ds^2 = -(1 - 2GM/c^2r + G^2J^2/c^4r^2) \, c^2dt^2 + \Sigma(1 + G^2J^2/c^4r^2\sin^2\theta)^{-1}dr^2 + \Sigma d\theta^2 + (r^2 + G^2J^2/c^4r^2\sin^2\theta)\sin^2\theta \, d\varphi^2 - 4GJ/c^2r \sin^2\theta \, dt \, d\varphi$

Here, J is the angular momentum of the black hole, and $\Sigma = r^2 + (GJ/c^2)^2\cos^2\theta$.

The Cosmic Censorship Hypothesis is related to the properties of the Schwarzschild and Kerr metrics, and whether singularities (which arise when certain metric components diverge) can be observed by distant observers.

1. Penrose Inequality:

The Penrose Inequality is a conjecture related to the Cosmic Censorship Hypothesis that states:

$A \geq 16\pi \, (m\_irr)^2$

Here, A is the surface area of the black hole's event horizon, and m_irr is the irreducible mass of the black hole, which is related to the black hole's mass (M) and angular momentum (J) by the equation:

$m\_irr = (1/2) \, \sqrt{(M^2 - J^2/M^2)}$

The Penrose Inequality, if proven, would provide support for the Weak Cosmic Censorship Hypothesis.

While the Cosmic Censorship Hypothesis does not have specific equations associated with it, these fundamental equations and concepts in general relativity are closely related to the conjecture and the study of black holes and singularities.

The Cosmic Censorship Hypothesis represents a fundamental question about the structure and limits of spacetime in our universe. While the hypothesis has withstood many challenges over the years, its ultimate validity remains an open question that continues to inspire research and debate. The resolution of this conjecture could profoundly impact our understanding of black holes, singularities, and the nature of spacetime itself, paving the way for new insights into the deepest mysteries of the cosmos.

39

The Multiverse Theory

The multiverse theory, also known as the many-worlds interpretation, is a hypothesis in physics and philosophy that proposes the existence of multiple parallel universes, each with its own set of physical laws and properties. This theory has gained significant attention in recent years, as it has the potential to answer some of the most fundamental questions about the nature of reality.

The Origins of the Multiverse Theory

The concept of the multiverse is rooted in the field of quantum mechanics, which describes the behavior of particles at the smallest scales of existence. According to quantum mechanics, particles exist in a state of superposition, meaning that they can exist in multiple states simultaneously. It is only when a measurement is made that the particle's state collapses into a single, definite state.

In 1957, physicist Hugh Everett III proposed the many-worlds interpretation of quantum mechanics, which suggests that when a particle collapses into a definite state, the universe splits into multiple parallel universes, each corresponding to a different possible outcome of the measurement. This theory was initially controversial, as it challenged the traditional Copenhagen interpretation of quantum mechanics, which suggests that particles do not exist in definite states until they are measured.

Key Concepts of the Multiverse Theory

The multiverse theory proposes that there are an infinite number of parallel universes, each with its own set of physical laws and properties. Some of the key concepts of the multiverse theory include:

1. The idea of parallel universes - According to the multiverse theory, each parallel universe is a separate entity, with its own set of physical laws and properties. These universes are thought to be separate from one another, existing in a separate space-time continuum.

2. The concept of the multiverse as a solution to the fine-tuning problem - The multiverse theory has been proposed as a solution to the so-called fine-tuning problem, which refers to the fact that many of the physical constants and parameters of the universe appear to be finely tuned to support life. The multiverse theory suggests that there are an infinite number of universes with different physical properties, and that our universe is one of the few that happens to be finely tuned for life.

3. The role of observation in the multiverse - The multiverse theory suggests that the act of observation can have a profound effect on the universe. According to the theory, the act of observation can collapse the wave function of a particle, causing the universe to split into multiple parallel universes, each corresponding to a different possible outcome of the observation.

4. The possibility of interaction between parallel universes - Some versions of the multiverse theory propose that it may be possible for parallel universes to interact with one another, although the nature of such interactions is currently unknown.

Implications of the Multiverse Theory

The multiverse theory has significant implications for our understanding of the nature of reality. Some of the key implications include:

1. The possibility of infinite variations of reality - The multiverse theory suggests that there may be an infinite number of variations of reality, each with its own set of physical laws and properties. This has profound implications for our understanding of the universe, as it suggests that

the laws of physics that we observe in our universe may not be universal.

2. The potential resolution of the fine-tuning problem - The multiverse theory has been proposed as a solution to the fine-tuning problem, as it suggests that our universe is just one of many with different physical properties. This may help to explain why our universe appears to be so finely tuned for life.

3. The possibility of new scientific discoveries - The multiverse theory has the potential to open up new avenues of scientific research, as it suggests that there may be other universes with different physical properties and phenomena that we have yet to discover.

Philosophical implications - The multiverse theory also has significant philosophical implications. It challenges traditional notions of reality and raises questions about the nature of existence, consciousness, and free will.

Critiques of the Multiverse Theory

The multiverse theory has been the subject of significant debate and critique. Some of the key criticisms include:

1. Lack of empirical evidence - As of yet, there is no empirical evidence to support the existence of parallel universes. While the theory is mathematically consistent, some scientists argue that it is not testable and therefore cannot be considered a scientific theory.

2. Occam's razor - Occam's razor is a principle that states that, all things being equal, the simplest explanation is usually the best. Some critics argue that the multiverse theory violates this principle, as it posits the existence of an infinite number of universes, which seems unnecessarily complex.

3. The anthropic principle - The anthropic principle is the idea that the universe appears to be finely tuned for life because we are here to observe it. Some critics argue that the multiverse theory is a form of circular reasoning, as it uses the existence of life in our universe to support the existence of an infinite number of parallel universes.

The multiverse theory is a hypothesis in physics and philosophy that proposes the existence of multiple parallel universes, each with its own set of physical laws and properties. While the theory is mathematically consistent and has the potential to answer some of the most fundamental questions about the nature of reality, it has also been subject to significant debate and critique. As our understanding of the universe continues to evolve, the multiverse theory will likely remain an important topic of discussion in the scientific and philosophical communities.

The multiverse theory is a concept in physics and philosophy that posits the existence of multiple parallel universes, each with its own set of physical laws and properties. While the theory is largely theoretical and speculative, it has inspired numerous mathematical and theoretical frameworks. Some of the key equations and concepts related to the multiverse theory include:

1. Schrödinger equation - The Schrödinger equation is a fundamental equation in quantum mechanics that describes the behavior of particles at the smallest scales of existence. It is used to calculate the probability of finding a particle in a given location or state. The equation is given by:

$i\hbar(d/dt)\psi = H\psi$

where $\hbar$ is the reduced Planck constant, t is time, ψ is the wave function of the particle, and H is the Hamiltonian operator, which describes the total energy of the particle.

1. Wave function collapse - The wave function collapse is a key concept in quantum mechanics, which refers to the act of measurement causing the collapse of the particle's wave function into a single, definite state. The probability of the particle being in a particular state is given by the squared magnitude of the wave function. After measurement, the wave function collapses to the measured state.

2. Many-worlds interpretation - The many-worlds interpretation is a theory that posits the existence of multiple parallel universes, each corresponding to a different possible outcome of a quantum measurement. The theory suggests that each universe is a separate entity with its own set of physical laws and properties.

3. Everett equation - The Everett equation, proposed by Hugh Everett III in 1957, is a key equation in the many-worlds interpretation. The equation describes the evolution of the wave function of the universe over time, and shows how the wave function splits into multiple branches corresponding to different possible outcomes of a measurement.

4. The cosmological constant - The cosmological constant is a term in the Friedmann equations, which describe the evolution of the universe over time. The cosmological constant is thought to represent the energy of empty space, and is related to the acceleration of the expansion of the universe. The value of the cosmological constant has implications for the existence of parallel universes, as a large cosmological constant would make the universe more likely to be infinite, allowing for the possibility of an infinite number of parallel universes.

5. The fine-tuning problem - The fine-tuning problem refers to the apparent coincidence that many of the physical constants and parameters of the universe appear to be finely tuned to support life. The multiverse theory has been proposed as a solution to the fine-tuning problem, as it suggests that there may be an infinite number of universes with different physical properties, and that our universe happens to be one of the few that is finely tuned for life.

Overall, the multiverse theory is a complex and multifaceted concept that involves numerous equations and mathematical frameworks. While the theory remains largely speculative, it has the potential to answer some of the most fundamental questions about the nature of reality, and will likely continue to be a topic of discussion and debate in the scientific and philosophical communities.

40

The Ultimate Fate of the Universe: Big Freeze, Big Crunch, or Big Rip?

The ultimate fate of the universe is a topic of much debate and speculation in the field of cosmology. There are several competing theories about how the universe will end, each with its own set of predictions and implications for the future of the cosmos. Some of the most prominent theories include the Big Freeze, Big Crunch, and Big Rip scenarios.

The Big Freeze

The Big Freeze, also known as the Heat Death, is a theory that suggests the universe will continue to expand and cool over time, eventually reaching a state of maximum entropy where all matter is evenly distributed and no more energy can be extracted. At this point, all stars will have exhausted their fuel and all black holes will have evaporated due to Hawking radiation, leaving only cold, dark matter and radiation. In this scenario, the universe will become a vast, empty void, with no life or activity.

The Big Crunch

The Big Crunch theory suggests that the universe will eventually stop expanding and start contracting, eventually collapsing back in on itself in a massive implosion known as the Big Crunch. This scenario is based on the assumption that the universe contains enough matter to cause the expansion to slow down and eventually reverse. In this scenario, the universe would collapse in on itself, resulting in a massive explosion that would destroy everything in the universe.

The Big Rip

The Big Rip scenario is a more recent theory that suggests the universe will continue to expand at an accelerating rate, eventually becoming so large that the gravitational forces holding galaxies and other structures together will be overcome. This will result in a rapid and catastrophic expansion, known as the Big Rip, that will tear everything in the universe apart, including atoms and subatomic particles. In this scenario, the universe would be destroyed in a matter of moments, leaving nothing behind.

Implications and Evidence

Each of these theories has significant implications for our understanding of the universe and its future. The Big Freeze suggests that the universe will continue to expand and cool, eventually leading to a state of maximum entropy where no more energy can be extracted. This scenario is supported by observations of the cosmic microwave background radiation, which suggest that the universe is cooling and becoming more diffuse over time.

The Big Crunch scenario is based on the assumption that the universe contains enough matter to cause the expansion to slow down and eventually reverse. However, recent observations of the cosmic microwave background radiation and the distribution of galaxies suggest that the universe may not contain enough matter to cause a Big Crunch. Instead, the universe may continue to expand indefinitely.

The Big Rip scenario is based on the assumption that the expansion of the universe is accelerating, as suggested by observations of distant supernovae and the cosmic microwave background radiation. However, there is still much debate and uncertainty about the nature of dark energy, the mysterious force that is thought to be driving the acceleration of the universe.

Each of the three scenarios for the ultimate fate of the universe has mathematical implications and is based on our understanding of fundamental physics and cosmology. Some of the key equations related to these scenarios include:

1. The Friedmann equations - The Friedmann equations describe the expansion of the universe over time and are the foundation of modern cosmology. They are given by:

(1) $H^2 = (8\pi G/3)\rho - (kc^2/a^2) + \Lambda/3$

(2) $(dH/dt) = -(4\pi G/3)(\rho + 3P/c^2) + \Lambda/3$

where H is the Hubble parameter, G is the gravitational constant, ρ is the density of matter and radiation in the universe, P is the pressure of matter and radiation, c is the speed of light, a is the scale factor of the universe, k is the curvature of space, and Λ is the cosmological constant.

2.The second law of thermodynamics - The second law of thermodynamics states that the total entropy of a closed system will always increase over time. In the context of the Big Freeze scenario, this law implies that the universe will eventually reach a state of maximum entropy where all matter is evenly distributed and no more energy can be extracted. The entropy of the universe is given by:

(3) $S = k\_B \ln(W)$

where k_B is the Boltzmann constant and W is the number of microstates corresponding to a particular macrostate.

3.The acceleration equation - The acceleration equation describes the rate at which the expansion of the universe is accelerating, and is related to the dark energy that is thought to be driving this acceleration. The equation is given by:

(4) $(d^2a/dt^2) = -(4\pi G/3)\rho a + \Lambda a$

where a is the scale factor of the universe, ρ is the density of matter and radiation, and Λ is the cosmological constant.

4.The density parameter - The density parameter is a dimensionless quantity that describes the relative densities of matter and radiation in the universe. It is given by:

(5) $\Omega = \rho/\rho\_crit$

where ρ_crit is the critical density of the universe, defined as the density required for the universe to be flat. The value of the density parameter has implications for the ultimate fate of the universe, as a high enough value of Ω would cause the expansion of the universe to slow down and eventually reverse, leading to a Big Crunch.

Overall, the ultimate fate of the universe is a complex and multifaceted topic that involves numerous equations and mathematical frameworks. While the specific scenario for the end of the universe remains uncertain, our understanding of fundamental physics and cosmology provides important insights into the possible outcomes of the cosmos.

The ultimate fate of the universe is a topic of much debate and speculation in the field of cosmology. While there are several competing theories about how the universe will end, including the Big Freeze, Big Crunch, and Big Rip scenarios, there is still much uncertainty and debate about the nature of the universe and its future. As our understanding of the cosmos continues to evolve, it is likely that new theories and models will emerge, providing new insights into the ultimate fate of the universe.

IX

The Arrow of Time

41

Introduction to the Arrow of Time

The arrow of time is a concept in physics and philosophy that refers to the unidirectional nature of time, and the fact that time appears to flow in one direction, from past to future. This concept is closely related to the second law of thermodynamics, which states that the total entropy of a closed system will always increase over time. In this article, we will explore the concept of the arrow of time in more detail and its implications for our understanding of the universe.

The Arrow of Time in Physics

The arrow of time is a fundamental concept in physics, with implications for a wide range of fields, including thermodynamics, cosmology, and quantum mechanics. The arrow of time is closely related to the second law of thermodynamics, which states that the total entropy of a closed system will always increase over time. Entropy is a measure of the disorder or randomness of a system, and the second law of thermodynamics suggests that as time progresses, systems become more disordered and random.

One of the key implications of the arrow of time is that it provides a directionality to time that is not present in many other physical phenomena. For example, in classical mechanics, the equations of motion are time-reversible, meaning that they can be run forward or backward in time with the same results. However, in the context of the arrow of time, it is clear that time has a distinct directionality, with events moving from past to future.

Another important concept related to the arrow of time is the concept of causality. Causality suggests that events in the past cause events in the future, and that the directionality of time is intimately tied to this causal relationship. For example, we can only remember events that have already occurred in the past, not events that will occur in the future.

The Arrow of Time and Cosmology

The arrow of time is also closely related to our understanding of the origins and evolution of the universe. One of the key questions in cosmology is why the universe appears to be ordered and structured, despite the fact that the second law of thermodynamics suggests that entropy should be increasing over time.

One possible explanation for this is the concept of the early universe being in a state of low entropy. According to this theory, the universe began in a state of extreme order and structure, and as time progressed, the universe evolved toward a state of higher entropy. This theory is supported by observations of the cosmic microwave background radiation, which suggest that the early universe was homogeneous and isotropic.

Another possible explanation for the arrow of time in cosmology is the concept of inflationary cosmology. Inflationary cosmology suggests that the universe underwent a period of rapid expansion in the moments after the Big Bang, leading to the formation of large-scale structures and patterns in the universe. This theory is supported by observations of the large-scale structure of the universe, which suggest that there are distinct patterns and structures that are consistent with inflationary cosmology.

The Arrow of Time and Philosophy

The concept of the arrow of time has important implications for philosophy, particularly in the areas of metaphysics and epistemology. One of the key questions in metaphysics is the nature of time itself, and whether time is a real and objective feature of the universe, or whether it is a human construct.

The arrow of time also has important implications for our understanding of causality and the relationship between cause and effect. If time is unidirectional and moves from past to future, then it follows that causality must also be unidirectional, with events in the past causing events in the future.

The arrow of time is a fundamental concept in physics and philosophy that has important implications for our understanding of the universe and our place in it. While the exact nature of the arrow of time remains a topic of debate and The concept of the arrow of time has important implications for our understanding of the universe and our place in it, and there are several key equations that help to describe the phenomenon. Some of the most important equations related to the arrow of time include:

1. The second law of thermodynamics - The second law of thermodynamics states that the total entropy of a closed system will always increase over time. This law is expressed mathematically by the following equation:

(1) $\Delta S \geq 0$

where ΔS is the change in entropy over time.

1. The equation of motion in classical mechanics - The equation of motion in classical mechanics is a fundamental equation that describes the motion of objects under the influence of forces. The equation is given by:

(2) $F = ma$

where F is the force acting on the object, m is the mass of the object, and a is the acceleration of the object.

1. The equation of state in thermodynamics - The equation of state in thermodynamics describes the relationship between the pressure, volume, and temperature of a system. The ideal gas law is a well-known example of an equation of state, given by:

(3) $PV = nRT$

where P is the pressure of the gas, V is the volume of the gas, n is the number of moles of gas, R is the gas constant, and T is the temperature of the gas.

1. The Einstein field equations - The Einstein field equations describe the relationship between the curvature of space-time and the matter and energy present in the universe. These equations are given by:

(4) $G\mu\nu = (8\pi G/c^4)T\mu\nu$

where $G\mu\nu$ is the Einstein tensor, $T\mu\nu$ is the stress-energy tensor, G is the gravitational constant, and c is the speed of light.

1. The Schrödinger equation - The Schrödinger equation is a fundamental equation in quantum mechanics that describes the evolution of the wave function of a quantum system over time. The equation is given by:

(5) $i\hbar\ \partial\psi/\partial t = H\psi$

where $\hbar$ is the reduced Planck constant, ψ is the wave function of the system, t is time, and H is the Hamiltonian operator of the system.

Overall, the concept of the arrow of time is a complex and multifaceted topic that involves numerous equations and mathematical frameworks. While the specific details of the arrow of time remain a topic of debate and investigation, our understanding of fundamental physics and cosmology provides important insights into the nature of time and the evolution of the universe over time.

42

Time's Asymmetry and Entropy

Time's asymmetry and the concept of entropy are deeply interconnected, with both playing a critical role in our understanding of the universe and the nature of time itself. The asymmetry of time is a fundamental aspect of our experience, as we perceive events as occurring in a definite order, with the past distinctly different from the future. This apparent irreversibility of time has puzzled scientists and philosophers for centuries. In this article, we will explore the relationship between time's asymmetry and entropy, and how these concepts shape our understanding of the universe's evolution.

The Arrow of Time

The "arrow of time" is a concept that refers to the directionality or asymmetry of time, which distinguishes the past from the future. In classical mechanics and most other branches of physics, the laws are time-reversible, meaning that they remain valid when time is reversed. However, our everyday experience and many natural phenomena, such as the growth of plants, aging, and the cooling of hot objects, seem to indicate a preferred direction for the flow of time.

The Second Law of Thermodynamics

One of the most fundamental expressions of time's asymmetry is found in the Second Law of Thermodynamics, which states that the total entropy of an isolated system can never decrease over time. Entropy is a measure of the disorder or randomness in a system, and the Second Law implies that natural processes tend to increase the overall entropy of the universe. This law provides a thermodynamic arrow of time, as it dictates a preferred direction for the flow of time based on the increase of entropy.

Entropy and Time's Asymmetry

The connection between entropy and time's asymmetry can be traced back to the work of Austrian physicist Ludwig Boltzmann, who developed the statistical mechanics interpretation of entropy. Boltzmann showed that the macroscopic behavior of a system, such as its temperature and pressure, can be explained by the statistical properties of its microscopic constituents, such as atoms and molecules.

Boltzmann's statistical interpretation of entropy provides a deep connection between the microscopic and macroscopic aspects of the world. The entropy of a system is related to the number of microstates (specific arrangements of particles) that correspond to a given macrostate (a particular set of macroscopic properties). A high-entropy macrostate has a large number of possible microstates, making it more probable than a low-entropy macrostate with fewer microstates.

Boltzmann's interpretation of entropy suggests that the observed increase in entropy over time is a consequence of the statistical behavior of a vast number of particles in a system. As time progresses, systems tend to evolve from less probable, low-entropy states to more probable, high-entropy states. This statistical tendency gives rise to the apparent irreversibility of time and many natural processes.

The Role of Initial Conditions: The Past Hypothesis

An essential aspect of understanding time's asymmetry and the connection to entropy is the role of initial conditions. The so-called "Past Hypothesis" posits that the early universe started in an extremely low-entropy state, near the Big Bang. This low-entropy state provided a unique initial condition that set the stage for the observed increase in entropy over time.

The Past Hypothesis suggests that the asymmetry of time is not a fundamental property of the laws of physics, but rather a consequence of the particular initial conditions of our universe. If the universe had started in a higher-entropy state, the observed arrow of time might have been different, or even absent altogether. This hypothesis highlights the importance of understanding the origins and nature of the early universe to fully grasp the relationship between time's asymmetry and entropy.

Entropy and the Evolution of the Universe The increase in entropy over time has profound implications for the evolution of the universe and the emergence of complex structures, such as galaxies, stars, planets, and life. The low-entropy state of the early universe allowed for the formation of these structures through a series of processes driven by gravitational and nuclear forces. As the universe expanded and cooled, matter clumped together due to gravity, eventually forming galaxies and stars. Inside stars, nuclear fusion processes generated heavier elements, which were later dispersed into space through supernovae explosions, providing the building blocks for planets and life.

Cosmological theories, such as the ΛCDM model and the inflationary paradigm, provide a framework for understanding the evolution of the universe from its early low-entropy state to the current high-entropy state, characterized by the accelerated expansion driven by dark energy. These models predict that the universe will continue to expand and cool, with entropy increasing over time, leading to a future where stars exhaust their nuclear fuel, galaxies become increasingly isolated, and the universe approaches a state of maximum entropy, known as "heat death."

Quantum Mechanics and Time's Asymmetry

The relationship between entropy and time's asymmetry becomes even more intriguing when considering quantum mechanics. In quantum mechanics, the time evolution of a system is described by the Schrödinger equation, which is time-reversible. However, the process of quantum measurement, through which we obtain information about a system, appears to introduce an irreversible, time-asymmetric element.

The phenomenon of wavefunction collapse, in which a quantum system transitions from a superposition of states to a definite state upon measurement, seems to break the time-reversal symmetry of quantum mechanics. This apparent conflict has led to various interpretations of quantum mechanics, such as the many-worlds interpretation and the de

Broglie-Bohm theory, which attempt to reconcile the time-reversible nature of the Schrödinger equation with the seemingly irreversible process of measurement.

Entropy, Information, and Time's Asymmetry

The concept of entropy is also intimately connected to information theory, as demonstrated by Claude Shannon's development of the mathematical framework for quantifying information in terms of entropy. In this context, entropy can be interpreted as a measure of the uncertainty or missing information about a system.

The connection between entropy and information provides another perspective on time's asymmetry. As systems evolve and entropy increases, information is effectively lost or dissipated, leading to the observed irreversibility of natural processes. This perspective has led to the development of the field of nonequilibrium thermodynamics, which explores the interplay between entropy, information, and the irreversible dynamics of systems driven away from equilibrium.

Time's asymmetry and the concept of entropy are deeply intertwined, shaping our understanding of the universe and the nature of time itself. The increase of entropy over time, as dictated by the Second Law of Thermodynamics, provides a thermodynamic arrow of time that is consistent with our everyday experience and the evolution of the universe. The statistical mechanics interpretation of entropy, the role of initial conditions, and the connections to quantum mechanics and information theory all contribute to our understanding of this fascinating relationship.

Exploring the irreversible nature of time and the role of entropy in shaping the universe's evolution continues to be a central theme in modern physics and cosmology. As we strive to uncover the mysteries of the universe's origins and ultimate fate, the interplay between time's asymmetry and entropy will undoubtedly play a crucial role in shaping our understanding of the cosmos and our place within it

43

The Thermodynamic Arrow of Time

The thermodynamic arrow of time is a concept in physics that refers to the directionality of physical processes in time, as dictated by the second law of thermodynamics. The second law of thermodynamics states that the total entropy of a closed system will always increase over time, leading to the appearance of an arrow of time that points in the direction of increasing entropy. In this article, we will explore the concept of the thermodynamic arrow of time in more detail, and its implications for our understanding of the universe.

The Second Law of Thermodynamics

The second law of thermodynamics is a fundamental law of physics that governs the behavior of energy and matter in the universe. The law states that the total entropy of a closed system will always increase over time, or at best remain constant in the idealized case of a reversible process.

Entropy is a measure of the disorder or randomness of a system, and is related to the number of ways in which the particles in a system can be arranged. As a result of the second law, the universe tends to move towards a state of maximum entropy, where energy is evenly distributed and no more work can be extracted.

Implications of the Thermodynamic Arrow of Time

The thermodynamic arrow of time has several important implications for our understanding of the universe. One of the key implications is that the arrow of time provides a directionality to physical processes that is not present in many other phenomena. For example, in classical mechanics, the equations of motion are time-reversible, meaning that they can be run forward or backward in time with the same results. However, in the context of the arrow of time, it is clear that time has a distinct directionality, with events moving from past to future.

Another important implication of the thermodynamic arrow of time is the relationship between the arrow of time and causality. Causality suggests that events in the past cause events in the future, and that the directionality of time is intimately tied to this causal relationship. For example, we can only remember events that have already occurred in the past, not events that will occur in the future.

The thermodynamic arrow of time also has important implications for our understanding of the origins and evolution of the universe. One of the key questions in cosmology is why the universe appears to be ordered and structured, despite the fact that the second law of thermodynamics suggests that entropy should be increasing over time.

One possible explanation for this is the concept of the early universe being in a state of low entropy. According to this theory, the universe began in a state of extreme order and structure, and as time progressed, the universe evolved toward a state of higher entropy. This theory is supported by observations of the cosmic microwave background radiation, which suggest that the early universe was homogeneous and isotropic.

Another possible explanation for the thermodynamic arrow of time in cosmology is the concept of inflationary cosmology. Inflationary cosmology suggests that the universe underwent a period of rapid expansion in the moments after the Big Bang, leading to the formation of large-scale structures and patterns in the universe. This theory is supported by observations of the large-scale structure of the universe, which suggest that there are distinct patterns and structures that are consistent with inflationary cosmology.

The concept of the thermodynamic arrow of time is intimately connected with the second law of thermodynamics, which describes the directionality of physical processes over time. Some of the key equations related to the thermodynamic arrow of time include:

1. The Second Law of Thermodynamics: The second law of thermodynamics can be expressed mathematically as:

$dS \geq 0$

where dS is the change in entropy over time. This equation describes the fact that the total entropy of a closed system will always increase over time or remain constant in the idealized case of a reversible process.

1. Entropy and Disorder: Entropy can be thought of as a measure of disorder or randomness in a system. Mathematically, the entropy of a system is given by:

$S = k\_B \ln W$

where S is the entropy, k_B is the Boltzmann constant, and W is the number of microstates that correspond to a given macrostate.

1. Boltzmann's Equation: Boltzmann's equation describes the time evolution of the probability distribution function for the positions and momenta of particles in a gas. The equation is given by:

$\partial f/\partial t + v \cdot \nabla f + F/m \cdot \nabla\_v f = C[f]$

where f is the distribution function, v is the velocity of the particles, F is the force acting on the particles, m is the mass of the particles, and C[f] is the collision operator.

1. The Heat Equation: The heat equation describes the flow of heat through a material over time. The equation is given by:

$\partial u/\partial t = k \nabla^2 u$

where u is the temperature of the material, t is time, k is the thermal conductivity of the material, and ∇^2 is the Laplace operator.

1. The Navier-Stokes Equations: The Navier-Stokes equations describe the motion of fluids over time. The equations are given by:

$$\rho\,(\partial v/\partial t + v \cdot \nabla v) = -\nabla p + \mu\,\nabla^2 v + f$$
$$\nabla \cdot v = 0$$

where ρ is the density of the fluid, v is the velocity of the fluid, p is the pressure of the fluid, μ is the viscosity of the fluid, and f is any external forces acting on the fluid.

Overall, the thermodynamic arrow of time is a complex and multifaceted topic that involves numerous equations and mathematical frameworks. While the specific details of the arrow of time remain a topic of debate and investigation, our understanding of fundamental physics and thermodynamics provides important insights into the nature of time and the evolution of the universe over time.

The thermodynamic arrow of time is a fundamental concept in physics that has important implications for our understanding of the universe and our place in it. While the exact nature of the arrow of time remains a topic of debate and investigation, our understanding of fundamental physics and cosmology provides important insights into the nature of time and the evolution of the universe over time.

44

The Quantum Arrow of Time

The quantum arrow of time is a concept in physics that describes the directionality of quantum processes in time. It is based on the principles of quantum mechanics, which describe the behavior of particles and energy at the quantum level. In this article, we will explore the concept of the quantum arrow of time in more detail, and its implications for our understanding of the universe.

Quantum Mechanics and Time

Quantum mechanics is a fundamental theory in physics that describes the behavior of matter and energy at the smallest scales. In the context of quantum mechanics, time is treated as a fundamental parameter, along with position and momentum.

One of the key features of quantum mechanics is that the behavior of particles and energy at the quantum level is probabilistic rather than deterministic. In other words, the outcome of a measurement or observation of a quantum system cannot be predicted with certainty, but rather only with a certain probability.

The Quantum Arrow of Time

The concept of the quantum arrow of time arises from the fact that certain processes in quantum mechanics are irreversible, meaning that they can only occur in one direction in time. One example of such a process is the process of measurement, where the quantum state of a system is collapsed into a single observable state.

In the context of quantum mechanics, the arrow of time can be thought of as pointing in the direction of increasing entropy, just as in classical thermodynamics. However, unlike in classical thermodynamics, the arrow of time in quantum mechanics is not necessarily tied to the direction of increasing disorder, but rather to the direction of irreversible processes.

Implications of the Quantum Arrow of Time

The concept of the quantum arrow of time has several important implications for our understanding of the universe. One of the key implications is that it provides a directionality to quantum processes that is not present in classical mechanics. This means that quantum processes have a distinct directionality in time, with events moving from past to future.

Another important implication of the quantum arrow of time is that it is intimately tied to the concept of measurement in quantum mechanics. The process of measurement is irreversible and collapses the quantum state of a system into a single observable state, which creates an arrow of time that is intimately tied to the process of observation.

The quantum arrow of time also has important implications for our understanding of the origins and evolution of the universe. In particular, it suggests that the early universe was in a state of low entropy, much like in the case of classical thermodynamics. This state of low entropy gave rise to the formation of structure and complexity in the universe, and has important implications for our understanding of the origins of life and the evolution of the universe over time.

The concept of the quantum arrow of time is a fundamental concept in physics that has important implications for our understanding of the universe and our place in it. While the exact nature of the quantum arrow of time remains a topic of debate and investigation, our understanding of quantum mechanics and the principles of irreversibility and entropy provide important insights into the nature of time and the evolution of the universe over time.

45

The Psychological and Cosmological Arrows of Time

The psychological and cosmological arrows of time are concepts in psychology and cosmology that describe the subjective and objective experiences of time. In this article, we will explore the concept of the psychological and cosmological arrows of time in more detail, and their implications for our understanding of the universe and human experience.

The Psychological Arrow of Time

The psychological arrow of time is a concept in psychology that describes the subjective experience of the flow of time. It refers to the fact that people experience time as moving forward in a particular direction, from past to present to future.

The psychological arrow of time is influenced by a variety of factors, including memory, attention, perception, and emotion. For example, people often experience time as moving more quickly when they are engaged in an enjoyable activity, and more slowly when they are bored or anxious.

The psychological arrow of time is subjective and can vary from person to person, but it is a universal experience that is shared by all humans. It is closely tied to our sense of self and our ability to plan and make decisions based on our past experiences and future expectations.

The Cosmological Arrow of Time

The cosmological arrow of time is a concept in cosmology that describes the objective directionality of physical processes in the universe. It is closely tied to the concept of entropy, which measures the disorder or randomness of a system.

The cosmological arrow of time is based on the second law of thermodynamics, which states that the entropy of a closed system will always increase over time, leading to the appearance of an arrow of time that points in the direction of increasing entropy. In other words, the universe tends to move towards a state of maximum entropy, where energy is evenly distributed and no more work can be extracted.

The cosmological arrow of time is irreversible and universal, and is intimately tied to our understanding of the origins and evolution of the universe. It is closely tied to the concept of the Big Bang, which is thought to have marked the beginning of the universe and the origin of time itself.

Implications of the Psychological and Cosmological Arrows of Time

The psychological and cosmological arrows of time have important implications for our understanding of the universe and human experience. The psychological arrow of time is intimately tied to our subjective experience of time, and has important implications for our sense of self and our ability to plan and make decisions based on our past experiences and future expectations.

The cosmological arrow of time, on the other hand, is objective and universal, and has important implications for our understanding of the origins and evolution of the universe. It is closely tied to the concept of entropy, which is a fundamental concept in physics and cosmology, and has important implications for our understanding of the ultimate fate of the universe.

Overall, the psychological and cosmological arrows of time are complex and multifaceted concepts that have important implications for our understanding of the universe and our place in it. While the specific details of these arrows of time remain topics of debate and investigation, our understanding of fundamental physics, psychology, and cosmology provides important insights into the nature of time and our experiences of it.

The psychological arrow of time is a subjective experience that is difficult to quantify mathematically. However, researchers have studied the neural mechanisms underlying the subjective experience of time, and have identified a number of brain regions and neural pathways that are involved in our perception of time.

One important model of the psychological arrow of time is the pacemaker-accumulator model. According to this model, the brain contains an internal clock that emits a regular pulse, or "pacemaker" signal. This signal is then accumulated over time by neural circuits that are responsible for timing and duration perception.

The pacemaker–accumulator model can be expressed mathematically using the following equation:

$$t = N \times \tau$$

where t is the perceived duration of a stimulus, N is the number of pulses emitted by the pacemaker during the stimulus, and τ is the duration of each pacemaker pulse.

The cosmological arrow of time, on the other hand, is based on the laws of physics and can be expressed mathematically using a number of equations from thermodynamics and cosmology.

One of the key equations related to the cosmological arrow of time is the second law of thermodynamics, which describes the directionality of physical processes over time. The second law of thermodynamics can be expressed mathematically as:

$$\Delta S \geq 0$$

where ΔS is the change in entropy over time. This equation describes the fact that the total entropy of a closed system will always increase over time or remain constant in the idealized case of a reversible process.

The cosmological arrow of time is also closely tied to the concept of entropy, which can be expressed mathematically using the following equation:

$$S = k_B \ln W$$

where S is the entropy, k_B is the Boltzmann constant, and W is the number of microstates that correspond to a given macrostate.

In cosmology, the cosmological arrow of time is closely tied to the expansion of the universe, which can be described mathematically using the Friedmann equations. The Friedmann equations describe the evolution of the universe over time, taking into account the effects of matter, radiation, and dark energy.

The Friedmann equations can be expressed mathematically as:

$(H/H\_0)\char`\^2 = \Omega\_m/a\char`\^3 + \Omega\_r/a\char`\^4 + \Omega\_\Lambda$

where H is the Hubble parameter, H_0 is the present-day value of the Hubble parameter, Ω_m is the density parameter for matter, Ω_r is the density parameter for radiation, Ω_Λ is the density parameter for dark energy, and a is the scale factor, which describes the expansion of the universe over time.

Overall, the psychological and cosmological arrows of time are complex concepts that are difficult to express mathematically. However, researchers have developed a number of mathematical models and equations that help to describe and quantify these arrows of time, and provide important insights into the nature of time and our experiences of it.

X

Wormholes and Time Travel

46

Introduction to Wormholes and Time Travel

Wormholes and time travel have captured the imagination of scientists, philosophers, and science fiction enthusiasts alike. They represent intriguing possibilities that challenge our understanding of spacetime and the fundamental laws of physics. While they may seem like the stuff of science fiction, wormholes and time travel have a basis in the rigorous mathematics of general relativity, the theory of gravitation developed by Albert Einstein. In this article, we will introduce the concepts of wormholes and time travel, explore their theoretical foundations, and discuss the ongoing scientific debate surrounding their feasibility.

Wormholes: Tunnels Through Spacetime

A wormhole, also known as an Einstein-Rosen bridge, is a hypothetical structure that connects two separate points in spacetime. It can be visualized as a tunnel-like shortcut that allows for faster-than-light travel between the two connected regions. The concept of a wormhole is derived from the mathematics of general relativity, which describes the curvature of spacetime in the presence of mass and energy.

Wormholes are solutions to the Einstein field equations that represent "nontrivial topologies" of spacetime. These solutions involve a seamless joining of two separate spacetime regions through a "throat" or tunnel. The most famous example of a wormhole solution is the Schwarzschild wormhole, which is derived from the Schwarzschild metric that describes the spacetime geometry around a non-rotating black hole. However, the Schwarzschild wormhole is inherently unstable and would collapse before any meaningful traversal could occur.

The concept of traversable wormholes, which would allow for the passage of matter and information between the connected regions, was later developed by physicists such as Kip Thorne and his collaborators. The existence of traversable wormholes would require exotic forms of matter with negative energy density, known as "negative energy" or "exotic matter," to prevent the wormhole from collapsing.

Time Travel: Paradoxes and Possibilities

Time travel is another fascinating concept that emerges from the mathematics of general relativity. In a relativistic framework, time is treated as a dimension similar to space, and the curvature of spacetime can lead to intriguing scenarios where the future and the past become connected. Closed timelike curves (CTCs) are one such example, where the worldlines of particles loop back on themselves, allowing for the possibility of time travel.

The existence of CTCs raises several paradoxes and challenges to our understanding of causality and the fundamental laws of physics. The most famous example is the "grandfather paradox," where a time traveler goes back in time and kills their own grandfather, preventing their own existence. This paradox highlights the potential inconsistencies that arise from the possibility of time travel and has led to various proposed resolutions, including the Novikov self-consistency principle, which posits that any events in a CTC must be self-consistent and that the probability of a paradox occurring is zero.

Wormholes and Time Travel: The Connection

The connection between wormholes and time travel arises from the fact that a traversable wormhole could potentially be used to create CTCs. If two ends of a wormhole were to be moved relative to each other or placed in regions with different gravitational potentials, they would experience different rates of time dilation, as predicted by general relativity. By carefully manipulating the wormhole's geometry, it may be possible to create a scenario where one end of the wormhole is connected to the past of the other end, effectively allowing for time travel.

Feasibility and Ongoing Research

The feasibility of wormholes and time travel remains an open question in the scientific community. The requirement of exotic matter with negative energy density to maintain a traversable wormhole is one of the major challenges, as such matter has not been observed in nature and its existence is still a matter of theoretical speculation. Moreover, the energy requirements for creating and stabilizing a wormhole are likely to be immense, possibly exceeding the energy available in the observable universe.

Another challenge comes from the potential violation of causality that time travel implies. The paradoxes associated with time travel raise concerns about the consistency of the laws of physics and the very nature of reality. Some physicists argue that the presence of time travel or CTCs would lead to unacceptable violations of causality, and therefore, the universe must have some mechanism to prevent their occurrence. This idea is known as the "chronology protection conjecture," proposed by Stephen Hawking.

In recent years, researchers have also begun exploring the implications of quantum mechanics for wormholes and time travel. Quantum mechanics introduces a new layer of complexity and may provide additional constraints on the feasibility of wormholes and CTCs. For instance, studies have suggested that quantum effects could lead to the "quantum backreaction" phenomenon, where the quantum fluctuations within a wormhole would destabilize it and prevent time travel.

Despite these challenges, the study of wormholes and time travel continues to be an active area of research, as they provide valuable insights into the nature of spacetime, the fundamental laws of physics, and the limits of our understanding of the universe. Ongoing research efforts are focused on refining the theoretical models of wormholes, exploring their possible astrophysical signatures, and investigating the potential interplay between general relativity and quantum mechanics in the context of wormholes and time travel.

Wormholes and time travel represent fascinating frontiers in our understanding of spacetime and the universe. Rooted in the mathematics of general relativity, these concepts push the boundaries of our understanding of physics and challenge our perception of reality. While the feasibility of wormholes and time travel remains an open question, their study continues to provide valuable insights into the nature of spacetime, the interplay between general relativity and quantum mechanics, and the ultimate limits of our understanding of the cosmos. As we continue to probe the frontiers of spacetime, the mysteries of wormholes and time travel will undoubtedly continue to captivate the minds of scientists and laypeople alike.

47

General Relativity and Wormholes

General relativity is a fundamental theory of physics that describes the behavior of gravity and the curvature of spacetime. It was developed by Albert Einstein in the early 20th century, and has since been confirmed by numerous experiments and observations.

One of the predictions of general relativity is the existence of wormholes, which are hypothetical structures that connect different regions of spacetime. In this article, we will explore the concept of general relativity and wormholes in more detail.

General Relativity

General relativity describes the behavior of gravity as the curvature of spacetime caused by the presence of matter and energy. According to general relativity, objects with mass and energy cause a curvature of spacetime, which affects the motion of other objects in the vicinity.

The curvature of spacetime is described by the Einstein field equations, which are a set of partial differential equations that relate the curvature of spacetime to the distribution of matter and energy. The Einstein field equations can be expressed mathematically as:

$G\mu\nu = 8\pi T\mu\nu$

where $G\mu\nu$ is the Einstein tensor, which describes the curvature of space-time, and $T\mu\nu$ is the stress-energy tensor, which describes the distribution of matter and energy.

Wormholes

A wormhole is a hypothetical structure that connects two different regions of spacetime, allowing for faster-than-light travel or communication. According to general relativity, wormholes are possible solutions to the Einstein field equations, although they require exotic matter with negative energy densities to stabilize them.

The geometry of a wormhole can be described by the Morris-Thorne metric, which is a solution to the Einstein field equations that describes a wormhole with a throat that connects two different regions of spacetime. The Morris-Thorne metric can be expressed mathematically as:

$$ds^2 = -e^{2\Phi(r)}dt^2 + dr^2/(1 - b(r)/r) + r^2(d\theta^2 + \sin^2\theta d\phi^2)$$

where $e^{2\Phi(r)}$ describes the redshift factor, $b(r)$ is the shape function that determines the shape of the wormhole, and r, θ, and ϕ are the radial and angular coordinates.

Implications of Wormholes

The existence of wormholes has important implications for our understanding of the universe and the possibility of interstellar travel. Wormholes could provide a way to travel faster than the speed of light, or to travel long distances through space without the need for conventional propulsion systems.

However, the existence of wormholes is still a topic of debate and investigation, and their stability and feasibility as a means of interstellar travel remain uncertain. The concept of wormholes also raises important questions about the nature of spacetime and the behavior of matter and energy at the quantum level, and has important implications for our understanding of the fundamental laws of physics.

The concept of wormholes is fascinating and raises important questions about the nature of spacetime and the laws of physics. One of the most intriguing aspects of wormholes is the possibility of time travel, which arises from the fact that a wormhole could connect two different points in spacetime.

According to general relativity, time travel is theoretically possible, although it requires exotic matter with negative energy densities and the construction of a stable wormhole. The mathematical equations that describe the behavior of a wormhole suggest that time travel could be possible if one end of the wormhole were accelerated to near the speed of light and then brought back to the starting point. This would create a "time machine" that would allow for travel to the past or the future.

However, the possibility of time travel raises important questions about causality and the nature of the universe. If time travel were possible, it could lead to paradoxes such as the grandfather paradox, in which a time traveler goes back in time and prevents their grandfather from meeting their grandmother, thus preventing their own existence.

Another intriguing aspect of wormholes is the potential for black hole wormholes, which would connect two different regions of spacetime through the interior of a black hole. Black hole wormholes are purely hypothetical, but they have important implications for our understanding of black holes and the behavior of matter and energy at the quantum level.

Overall, the concept of wormholes is a fascinating topic that raises important questions about the nature of spacetime and the laws of physics. While the feasibility and stability of wormholes as a means of interstellar travel remain uncertain, their existence provides important insights into the fundamental nature of the universe and the behavior of matter and energy.

General relativity is a fundamental theory of physics that describes the behavior of gravity and the curvature of spacetime. One of the predictions of general relativity is the existence of wormholes, which are hypothetical structures that connect different regions of spacetime. While the feasibility and stability of wormholes as a means of interstellar travel remain uncertain, the concept of wormholes provides important insights into the nature of spacetime and the behavior of matter and energy at the quantum level.

48

Time Travel Paradoxes

Time travel is a fascinating concept that has captivated the human imagination for centuries. However, the possibility of time travel raises important questions and paradoxes that challenge our understanding of causality and the nature of the universe. In this article, we will explore some of the most famous time travel paradoxes.

· **Grandfather Paradox**

The grandfather paradox is one of the most famous time travel paradoxes. It arises from the question of what would happen if a time traveler were to go back in time and prevent their grandfather from meeting their grandmother. In this scenario, the time traveler would not be born, and therefore could not have gone back in time to prevent their grandfather from meeting their grandmother.

This paradox challenges the concept of causality, as it suggests that events in the past could be changed in a way that would make their own existence impossible. While there are a number of proposed solutions to the grandfather paradox, including the concept of multiple timelines, it remains a fundamental challenge to the possibility of time travel.

· **Predestination Paradox**

The predestination paradox is another famous time travel paradox that arises from the question of what would happen if a time traveler were to go back in time and inadvertently cause the events that led to their own existence. In this scenario, the time traveler's actions would be predetermined by their own future existence, and they would be unable to change the course of events.

This paradox challenges the concept of free will, as it suggests that events in the past could be predetermined by events in the future. While there are a number of proposed solutions to the predestination paradox, including the concept of causal loops, it remains a fundamental challenge to the possibility of time travel.

· Bootstrap Paradox

The bootstrap paradox is a time travel paradox that arises from the question of what would happen if a time traveler were to go back in time and introduce an object or idea that had no clear origin. In this scenario, the object or idea would be a "closed loop," with no clear beginning or end.

This paradox challenges the concept of causality, as it suggests that events in the past could be caused by events in the future, without any clear origin. While there are a number of proposed solutions to the bootstrap paradox, including the concept of parallel universes, it remains a fundamental challenge to the possibility of time travel.

Overall, the possibility of time travel raises important questions and paradoxes that challenge our understanding of the nature of the universe. While there are a number of proposed solutions to these paradoxes, they remain a fundamental challenge to the possibility of time travel, and provide important insights into the nature of causality and free will.

· Ontological Paradox

The ontological paradox is a time travel paradox that arises from the question of what would happen if a time traveler were to go back in time and become their own ancestor or creator. In this scenario, the time traveler's existence would be caused by their own actions, creating a "closed loop" with no clear origin.

This paradox challenges the concept of causality, as it suggests that events in the past could be caused by events in the future, without any clear origin. While there are a number of proposed solutions to the ontological paradox, including the concept of multiple timelines, it remains a fundamental challenge to the possibility of time travel.

· **Twin Paradox**

The twin paradox is a time travel paradox that arises from the question of what would happen if one twin were to travel through space at near-light speeds and return to Earth, while the other twin remained on Earth. In this scenario, the twin who traveled through space would experience time dilation, and would age more slowly than the twin who remained on Earth.

This paradox challenges the concept of time as a constant, and suggests that time is relative to the observer's motion and gravity. While the twin paradox has been experimentally confirmed through the use of atomic clocks, it remains a fundamental challenge to our understanding of the nature of time and the behavior of matter and energy.

The possibility of time travel raises important questions and paradoxes that challenge our understanding of the nature of the universe. While there are a number of proposed solutions to these paradoxes, they remain a fundamental challenge to the possibility of time travel, and provide important insights into the nature of causality, free will, and the behavior of matter and energy. The study of time travel and its paradoxes is a fascinating field of research that continues to captivate the imagination of scientists and the general public alike.

49

The Chronology Protection Conjecture

The Chronology Protection Conjecture is a hypothesis proposed by renowned theoretical physicist Stephen Hawking in 1991. The conjecture posits that the laws of physics conspire to prevent the formation of closed timelike curves (CTCs) or any other form of time travel that would lead to causal paradoxes. In essence, the Chronology Protection Conjecture suggests that the universe has a built-in mechanism to protect the consistency of causality and the flow of time.

Closed timelike curves are trajectories in spacetime that loop back on themselves, allowing an object or a particle to travel back in time and potentially interact with its past self. The existence of CTCs would create the possibility of time travel, which in turn raises numerous paradoxes and challenges to our understanding of causality, such as the famous "grandfather paradox."

The Chronology Protection Conjecture is not a proven theorem or a well-established principle in physics. Instead, it is a hypothesis that has emerged from the study of general relativity and the attempts to reconcile its predictions with our everyday experience and the observed behavior of the universe. The conjecture is based on the observation that many solutions to the Einstein field equations, which allow for CTCs or time travel, are either unstable or require the presence of exotic forms of matter with negative energy densities.

One of the key motivations behind the Chronology Protection Conjecture is the desire to preserve the consistency of the laws of physics and avoid the troubling implications of time travel paradoxes. If the universe were to allow for time travel and CTCs, it would challenge our understanding of causality and potentially undermine the predictive power of physical theories.

In recent years, the Chronology Protection Conjecture has been investigated in the context of both classical general relativity and quantum mechanics. Some studies have suggested that quantum effects, such as vacuum fluctuations and particle creation, could play a role in preventing the formation of CTCs or destabilizing any attempt to create time travel scenarios. These quantum effects may provide a natural mechanism for chronology protection, reinforcing the conjecture's validity.

However, it is important to note that the Chronology Protection Conjecture remains an open question in the scientific community. There is no definitive proof that the laws of physics inherently prevent time travel or the formation of CTCs. The conjecture serves as an intriguing hypothesis that guides ongoing research into the nature of spacetime, causality, and the limits of our understanding of the universe.

As the study of the Chronology Protection Conjecture continues, researchers are exploring various aspects of spacetime, causality, and the interplay between general relativity and quantum mechanics. The conjecture has inspired numerous theoretical investigations and has led to a deeper understanding of the complexities of time and the fabric of the universe.

One area of ongoing research is the examination of spacetime structures that could potentially support CTCs or time travel. Wormholes, for instance, are hypothetical tunnels connecting distant regions of spacetime that might enable time travel if their geometries were manipulated appropriately. The study of traversable wormholes, which would allow for the passage of matter and information, has revealed the challenges of creating and maintaining such structures, as they would likely require exotic forms of matter with negative energy densities to prevent collapse.

Another area of research related to the Chronology Protection Conjecture is the exploration of alternative theories of gravity and spacetime. These theories, such as Loop Quantum Gravity and String Theory, aim to provide a more fundamental description of the universe that unifies general relativity with quantum mechanics. By examining the implications of these alternative theories for the formation of CTCs and time travel, researchers hope to gain insights into the validity of the Chronology Protection Conjecture and the nature of spacetime itself.

Moreover, researchers are also studying the role of black holes and cosmic censorship in the context of chronology protection. Cosmic censorship is a conjecture that suggests that singularities—points of infinite density that are predicted to exist within black holes—must always be hidden from external observers behind an event horizon. This idea has implications for the Chronology Protection Conjecture, as the presence of naked singularities could potentially lead to the formation of CTCs and time travel scenarios.

Ultimately, the Chronology Protection Conjecture remains an open question that motivates ongoing research in the fields of general relativity, quantum mechanics, and cosmology. While there is no definitive proof that the laws of physics inherently prevent time travel or the formation of CTCs, the conjecture serves as an important guidepost for exploring the nature of spacetime, causality, and the fundamental limits of our understanding of the universe. As we continue to probe the mysteries of time and the fabric of reality, the Chronology Protection Conjecture will undoubtedly remain a topic of intense interest and debate in the scientific community.

50

The Unification of Physics and the Quest for a Theory of Everything

The unification of physics is a quest to find a single theory that can explain all the fundamental forces of nature, including gravity, electromagnetism, the strong nuclear force, and the weak nuclear force. This theory is often referred to as a Theory of Everything (TOE), and it represents the ultimate goal of modern physics.

The need for a TOE arises from the fact that our current understanding of the universe is based on two separate and seemingly incompatible theories: quantum mechanics and general relativity. While quantum mechanics describes the behavior of matter and energy at the smallest scales, general relativity describes the behavior of matter and energy at the largest scales, such as the behavior of galaxies and the universe as a whole.

The challenge of unifying these two theories arises from the fact that they are based on fundamentally different principles. Quantum mechanics is based on the principle of uncertainty, which suggests that particles can exist in multiple states at the same time and that their properties are determined by probabilities. General relativity, on the other hand, is based on the principle of curved spacetime, which suggests that matter and energy warp the fabric of spacetime, creating the force of gravity.

One approach to unifying these two theories is string theory, which suggests that particles are not point-like objects but rather tiny strings that vibrate at different frequencies. String theory proposes that the universe is made up of 10 or 11 dimensions, with the extra dimensions being curled up and hidden from our view. While string theory has not yet been proven experimentally, it has the potential to unify quantum mechanics and general relativity and provide a TOE.

Another approach to unifying physics is through the concept of super-symmetry, which proposes that each particle in the universe has a partner particle with a different spin. Supersymmetry suggests that the universe is made up of both matter particles (such as electrons and quarks) and force-carrying particles (such as photons and gluons), which are related to each other through supersymmetry.

The search for a TOE has important implications for our understanding of the universe and the behavior of matter and energy. A TOE would provide a complete and consistent description of the universe, from the smallest particles to the largest structures, and would enable us to make accurate predictions about the behavior of matter and energy in any situation.

In addition to its scientific importance, the quest for a TOE also has important philosophical implications. The unification of physics would provide a deep understanding of the nature of reality and the fundamental principles that govern the universe. It would also challenge our preconceived notions of the universe and our place in it, and could potentially lead to new technologies and discoveries that we cannot even imagine today.

Overall, the unification of physics and the search for a TOE represents one of the greatest challenges in modern science. While we may never fully understand the nature of the universe, the quest for a TOE provides an exciting and thought-provoking journey that continues to captivate the imagination of scientists and the general public alike.

The quest for a Theory of Everything involves the unification of the four fundamental forces of nature: gravity, electromagnetism, the strong nuclear force, and the weak nuclear force. These forces are described by different equations, and the challenge of unifying them lies in finding a single equation that can describe all of them. Here are some of the key equations involved in the unification of physics:

1. Newton's law of gravity: $F = G(m_1m_2)/r^2$, where F is the force between two objects, G is the gravitational constant, m_1 and m_2 are the masses of the objects, and r is the distance between them. This equation describes the force of gravity between two objects and was the basis of classical mechanics.

2. Maxwell's equations: These are a set of four equations that describe the behavior of electric and magnetic fields. They are:

3. a. Gauss's law for electric fields: $\nabla \cdot E = \rho/\varepsilon_0$, where $\nabla \cdot E$ is the divergence of the electric field, ρ is the charge density, and ε_0 is the electric constant.

4. b. Gauss's law for magnetic fields: $\nabla \cdot B = 0$, where $\nabla \cdot B$ is the divergence of the magnetic field.

5. c. Faraday's law of electromagnetic induction: $\nabla \times E = -\partial B/\partial t$, where $\nabla \times E$ is the curl of the electric field, B is the magnetic field, and t is time.

6. d. Ampere's law with Maxwell's correction: $\nabla \times B = \mu_0(J + \varepsilon_0\partial E/\partial t)$, where $\nabla \times B$ is the curl of the magnetic field, μ_0 is the magnetic constant, J is the current density, and $\partial E/\partial t$ is the time derivative of the electric field.

7. The Schrödinger equation: $i\hbar(\partial\psi/\partial t) = H\psi$, where i is the imaginary unit, $\hbar$ is the reduced Planck constant, ψ is the wave function, t is time, and H is the Hamiltonian operator. This equation describes the behavior of particles in quantum mechanics.

8. Einstein's field equations: $G\mu\nu = 8\pi T\mu\nu$, where $G\mu\nu$ is the Einstein tensor, $T\mu\nu$ is the stress-energy tensor, and 8π is a constant. These equations describe the behavior of matter and energy in general relativity.

9. String theory equations: String theory proposes that particles are not point-like objects but rather tiny strings that vibrate at different frequencies. The equations involved in string theory are complex and involve advanced mathematical concepts, such as Riemann surfaces and conformal field theory. The key equation in string theory is the

worldsheet action, which describes the behavior of strings in spacetime.

While these equations describe different aspects of the universe, the challenge of unifying them lies in finding a single equation that can describe all of them. This equation is known as the Theory of Everything, and it remains one of the greatest challenges in modern physics.

Summary:

"Our Picture of the Universe" is a comprehensive exploration of humanity's understanding of the cosmos, from ancient times to modern scientific discoveries. The book begins with a historical overview of astronomical knowledge, from the geocentric model to the Copernican Revolution and Galileo's groundbreaking work with the telescope. The book then delves into the nature of space and time, discussing Isaac Newton's Principia Mathematica and Einstein's theory of relativity, along with their implications for time dilation and the geometry of spacetime.

The book also examines the expanding universe, discussing Edwin Hubble's redshift observations, the Big Bang Theory, and the mysteries of dark matter and dark energy. Quantum mechanics and the uncertainty principle are explored, including the principles of wave-particle duality, Heisenberg's uncertainty principle, and the various interpretations of quantum mechanics.

The book provides an in-depth look at elementary particles and the forces of nature, covering the Standard Model of particle physics, the four fundamental forces, and the Higgs boson. Black holes are discussed extensively, including their formation, structure, and the detection of gravitational waves from black hole mergers.

Stephen Hawking's work on black hole radiation is presented, along with the information paradox, black hole entropy, and the holographic principle. The book also investigates the origin and fate of the universe, including the inflationary universe, cosmic censorship, the multiverse theory, and possible ultimate fates such as the Big Freeze, Big Crunch, or Big Rip.

The concept of the arrow of time is explored, addressing time's asymmetry and entropy, the thermodynamic arrow of time, and the quantum and psychological arrows of time. Finally, the book delves into wormholes and time travel, covering general relativity, time travel paradoxes, and the chronology protection conjecture. Our Picture of the Universe provides a thorough examination of the past, present, and future of our understanding of the cosmos, ultimately highlighting the ongoing quest for a unified theory of everything.